T0187932

Plants of Central Asia

Volume 7

Plants of Central Asia

Plant Collections from China and Mongolia

Volume 7

Liliaceae – Orchidaceae

V.I. Grubov and T.V. Egorova

CRC Press
Taylor & Francis Group
Boca Raton London New York

CRC Press is an imprint of the
Taylor & Francis Group, an **informa** business
A SCIENCE PUBLISHERS BOOK

ACADEMIA SCIENTIARUM URSS
INSTITUTUM BOTANICUM nomine V.L. KOMAROVII
PLANTAE ASIAE CENTRALIS
(*secus materies Instituti botanici nomine V.L. Komarovii*)
Fasciculus 7
LILIACEAE — ORCHIDACEAE
Conficerunt : V.I. Grubov et T.V. Egorova

First published 2003 by Science Publishers Inc.

Published 2019 by CRC Press
Taylor & Francis Group
6000 Broken Sound Parkway NW, Suite 300
Boca Raton, FL 33487-2742

© 2003, Copyright reserved
CRC Press is an imprint of Taylor & Francis Group, an Informa business

First issued in paperback 2019

No claim to original U.S. Government works

ISBN 13: 978-0-367-44690-1 (pbk)
ISBN 13: 978-1-57808-118-9 (hbk)
ISBN 13: 978-1-57808-062-5 (Set)

This book contains information obtained from authentic and highly regarded sources. Reasonable efforts have been made to publish reliable data and information, but the author and publisher cannot assume responsibility for the validity of all materials or the consequences of their use. The authors and publishers have attempted to trace the copyright holders of all material reproduced in this publication and apologize to copyright holders if permission to publish in this form has not been obtained. If any copyright material has not been acknowledged please write and let us know so we may rectify in any future reprint.

Except as permitted under U.S. Copyright Law, no part of this book may be reprinted, reproduced, transmitted, or utilized in any form by any electronic, mechanical, or other means, now known or hereafter invented, including photocopying, microfilming, and recording, or in any information storage or retrieval system, without written permission from the publishers.

For permission to photocopy or use material electronically from this work, please access www.copyright.com (http://www.copyright.com/) or contact the Copyright Clearance Center, Inc. (CCC), 222 Rosewood Drive, Danvers, MA 01923, 978-750-8400. CCC is a not-for-profit organization that provides licenses and registration for a variety of users. For organizations that have been granted a photocopy license by the CCC, a separate system of payment has been arranged.

Trademark Notice: Product or corporate names may be trademarks or registered trademarks, and are used only for identification and explanation without intent to infringe.

Visit the Taylor & Francis Web site at
http://www.taylorandfrancis.com

and the CRC Press Web site at
http://www.crcpress.com

Library of Congress Cataloging-in-publication Data

Rasteniia Tsentral 'noĭ Azii. English
  Plants of Central Asia: plant collections from China
  and Mongolia
  / [editor-in-chief, V.I. Grubov].
     p. cm.
  Research based on the collections of the V.L.
  Komarov Botanical Institute.
  Includes bibliographical references.
  Contents: v. 7. Liliaceae - Orchidaceae
  ISBN 1-57808-118-1
     1. Botany -- Asia, Central. I. Grubov, V. I. II.
Botabicheskiĭ institut im. V.L. Komarova. III. Title.
QK374, R23613 2002
581.958--dc21   99-36729
                                          CIP

Translation of: Rasteniya Central'nov Asii, vol. 7, 1977;
                  Nauka Publishers, Leningrad.

ANNOTATION

UDC 582.57+582.59 (51)

PLANTS OF CENTRAL ASIA. From the Materials of the V.L. Komarov Botanical Institute, Academy of Sciences of the USSR. Vol. 7. Liliaceae—Orchidaceae. Compilers: V.I. Grubov and T.V. Egorova. 1977. Nauka, Leningrad.

This is the seventh volume of the illustrated enumeration of the Plants of Central Asia (within the People's Republics of China and Mongolia) and covers families Liliaceae to Orchidaceae inclusive. The description of monocotyledonous families (volumes 3, 4 and 6) concludes with this volume. In the system adopted, this volume should have followed the third (Cyperaceae—Juncaceae). Onions represent the largest and important group in this volume; many species of this genus are valuable fodder plants as well as characteristic plants of desert steppes. From the phytogeographic viewpoint, genus iris and genera of family Liliaceae are most interesting.

V.I. Grubov
Editor-in-Chief and Volume Editor

PREFACE

This, the seventh volume of Plants of Central Asia, contains treatments of families Liliaceae, Dioscoreaceae, Amaryllidaceae, Iridaceae and Orchidaceae, and concludes treatments of all monocotyledonous plants. The taxa of these Central Asian families covered in the present volume are as follows: Liliaceae 19 genera with 179 species, Orchidaceae 18 genera with 30 species, Iridaceae 2 genera with 21 species and Amaryllidaceae and Dioscoreaceae one species each. With inclusion of the species distributed in the Soviet territory of Central Asia, the additions are: for Liliaceae 2 genera with some 80 species, Orchidaceae one genus with 4 species and Iridaceae one genus with 4 species; thus, a total of 4 genera with about 90 species.

Two more genera of Liliaceae with 7 species, 2 species of Iridaceae and 3 species of Orchidaceae have also been included in the keys but are not numbered in the text. All of these are reported from immediately adjoining regions and the probability of their occurrence within non-Soviet Central Asia is not excluded. Thus, this volume covers 43 genera with 191 species.

The largest genera are: *Allium* L. with 81 species and *Iris* L. with 20 species. These two genera contain the maximum number of species endemic to Central Asia: among onions, 23 are endemic and 19 subendemic (i.e., more than half of all species) and among iris, 2 and 6 respectively. The species of these genera contribute significantly to the composition of the plant cover of Central Asia. Onions—*A. mongolicum, A. polyrhizum, A. bidentatum, A. odorum, A. oreoprasum*—represent important constituents of the coenosis. They comprise the basis of so-called onion steppes which cover vast areas in Mongolia and provide excellent pasture in the desert steppe subzone (semi-desert). Mongolian cattle breeders prepare from the green succulent onion mass of *A. mongolicum* and *A. polyrhizum* highly nutritious salted cakes for winter-feeding of cattle. Some iris species also enjoy a special place in the vegetation cover of Central Asia and even became landscape plants. *Iris bungei* is a characteristic plant and an important element (adherent) of arid and desert steppes of Central Mongolia and *I. tenuifolia* of sandy cattail steppes of Mongolia and Kazakhstan. *I. dichotoma* forms typical iris steppes (arid steppes) in easternmost Mongolia. *I. lactea* and *I. loczii* are highly prominent landscape plants, forming vast bright green beds and belts along rivers and brooks and around springs.

The groups of monocotyledonous families covered here is neither abundant nor diverse. Only some genera of lilies and irises have evolved into specific series and cycles of species within Central Asia. An example is *Allium* L., represented here by a few cycles of species which include numerous endemics characteristic of nearly each of the Central Asian regions. Other examples are *Gagea* Salisb. and *Tulipa* L., confined almost exclusively to Junggar with endemic species there. Mention should also be made of several widely distributed desert species of asparagus—*Asparagus angulofractus, A. breslerianus, A. gobicus.*

Genus *Iris* L. has its own Central Asian cycle of closely related montane steppe and high-altitude (alpine) species (*I. pandurata, I. potaninii, I. thoroldi, I. tigridia*), their general distribution range extending from Tibet to the south of Eastern Siberia through Mongolia, as well as many steppe and desert-steppe species—*I. bungei* and *I. ventricosa.*

The remaining higher monocotyledonous genera are represented in Central Asia by isolated links, entering this region from adjacent regions of their main distribution range. A few are represented here by endemic and subendemic species, some being relict neoendemics with very small distribution ranges. These include the following species, described here as new: *Lilium tianschanicum* sp. nova, *Listera tianschanica* sp. nova (Eastern Tien Shan) and *Asparagus przewalskii* sp. nova, as well as *Fritillaria przewalskii, Galeorchis roborovskii* (Qinghai). Only Junggar *Fritillaria pallidiflora, Eremurus altaicus* and *E. anisopterus* from the Middle Asian groups are of greater antiquity.

As mentioned above, nearly every region of Central Asia has its own endemic species of onions. Thus, *Allium leucocephalum, A. mongolicum, A. polyrhizum, A. stellerianum, A. anisopodum* are typical of Mongolia while *A. pevtzovii* is exclusive to Kashgar. Endemics for Junggar-Turan (Kazakhstan) province in general are: *A. coeruleum, A. delicatulum, A. galanthum, A. oliganthum, A. pallasii;* for Junggar: *A. glomeratum, A. karelinii, A. petraeum, A. plastyspathum, A. sairamense,*

A. semenovii, A. setifolium, A. subtilissimum, A. teretifolium; for Tien Shan exclusively: *A. juldusicolum, A. kaschianum, A. korolkovii, A. megalobulbon, A. tianschanicum, A. weschnjakowii*; exclusive to Junggar-Tarbagatai region: *A. robustum*. Typical for Tibet in general are: *A. chalcophengos, A. chrysocephalum, A. cyaneum, A. kansuense, A. platystylum, A. przewalskianum* and *A. tibeticum*, while *A. dentigerum* and *A. tanguticum* are exclusive to Qinghai.

Among other members of Liliaceae, mention should be made of endemic desert asparagus: *Asparagus gobicus* is characteristic of desert and semidesert Mongolian province and replaces there the related Kazakhstan-Fore Asian *A. breslerianus* as well as solonchak *A. angulofractus* whose range also extends into Junggar. Two genera of lilies, comprising endemic species, are almost wholly confined to Junggar: *Tulipa* L. and *Gagea* Salisb. With the exception of widely distributed Altay-Siberian montane tulip *Tulipa uniflora* found in north-western Mongolia, all others are found in Junggar or only in Tien Shan; among them, *T. altaica* (subendemic), *T. aristata, T. heterophylla, T. triphylla* are endemic in Junggar and *T. iliensis, T. kolpakowskiana, T. tetraphylla* and *T. tianschanica* in Tien Shan. The situation is similar in the case of genus *Gagea*. Only *G. granulosa* and *G. pauciflora* have comparatively broader distribution ranges while all others are confined to Junggar; further, *G. albertii* and *G. sacculifera* are endemic in it while *G. emarginata* represents a Tien Shan-Pamiro-Alay species. In fact, *Tulipa* as well as *Gagea* are Fore Asian genera, only entering Central Asia.

Endemic species of genus *Iris* L. likewise have their own individual distribution ranges. Thus, *I. bungei* is an exclusive plant of desert and semidesert Mongolia and *I. scariosa* a characteristic plant of rocky arid steppes and semi-deserts of Kazakhstan. *I. thoroldii* is an endemic species in alpine Tibet and *I. pandurata* a narrow endemic in Qinghai.

During field studies in Fore Hinggan in the course of the 1970 expedition, we were convinced that species of Manchurian flora predominate in this region even though the landscape is typically that of Hinggan (abundant mixed-grass montane meadows, birch groves and larch forests with undergrowths of wild rosemary, honeysuckle, spirea and spindle tree, whortleberry and mountain cranberry in the grass belt, sedge swamps, diverse coastal scrubs of hawthorn, bird cherry, alder, buckthorn, etc.) and thus this region falls in the Manchurian province of the East Asian subregion of the Holarctic.

Commencing with this volume, therefore, Fore Hinggan is excluded from Central Asia and will be mentioned only when referring to the general distribution of plants as a region of Northern Mongolia (since it figures in the phytogeographic zonation of the Mongolian People's Republic); for China, it is naturally included as a part of Dunbei and is not particularly cited (see map and scheme of phytogeographic division of Central Asia, facing p. 266 (map) and pp. 267, 268, 269.

In this volume, plates of plant drawings were prepared by artists G.M. Aduevskaya (Plate VI) and N.K. Voronkova (Plates I–V). Maps of distribution ranges were drawn by O.I. Starikova (Map 5) and I.B. Tikhmeneva (Maps 1–4). O.I. Starikova translated the text on labels of herbarium specimens of Chinese collectors (Sinkiang and Eastern Mongolia).

CONTENTS

TAXONOMY

SPECIAL ABBREVIATIONS

Abbreviations of Names of Collectors

A. Reg.	—	A. Regel
Bar.	—	V.I. Baranov
Chaff.	—	J. Chaffonjon
Chaney.	—	R.W. Chaney
Ching	—	R.C. Ching
Chu	—	C.N. Chu
Czet.	—	S.S. Czetyrkin
Divn.	—	D.A. Divnogorskaya
Fet.	—	A.M. Fetisov
Glag.	—	S.A. Glagolev
Gr.-Grzh.	—	G.E. Grum-Grzhimailo
Grombch.	—	B.L. Grombchevski
Grub.	—	V.I. Grubov
Gus.	—	V.A. Gusev
Ik.-Gal.	—	N.P. Ikonnikov-Galitzkij
Isach.	—	E.A. Isachenko
Ivan	—	A.F. Ivanov
Kal.	—	A.V. Kalinina
Klem.	—	E.N. Klements
Kom.	—	V.L. Komarov
Krasch	—	I.M. Krascheninnikov
Kryl.	—	P.N. Krylov
Kuan	—	K.S. Kuan
Lad.	—	V.F. Ladygin
Ladyzh.	—	M.V. Ladyzhinsky
Lis.	—	V.I. Lisovsky
Litw.	—	D.I. Litwinow
Lom.	—	A.M. Lomonossov
Merzb.	—	G. Merzbacher
Mois.	—	V.S. Moiseenko
Nov.	—	V.F. Novitski
Pal.	—	I.V. Palibin
Pavl.	—	N.V. Pavlov

Petr.   — M.P. Petrov
Pias.   — P.Ya. Piassezki
Pob.   — E.G. Pobedimova
Pop.   — M.G. Popov
Pot.   — G.N. Potanin
Przew.   — N.M. Przewalsky
Rachk.   — E.I. Rachkovskaya
A. Reg.   — A. Regel
Rgl.   — E.L. Regel
Rhins   — J.L. Dutreuil de Rhins
Rob.   — V.I. Roborowsky
Sap.   — V.V. Sapozhnikov
Schischk.   — B.K. Schischkin
Serp.   — V.M. Serpukhov
Sold.   — V.V. Soldatov
Tug.   — A.Ya. Tugarinov
Wang   — K.S. Wang
Yun.   — A.A. Yunatov
Zab.   — D.K. Zabolotnyi
Zam.   — B.M. Zamatkinov

Abbreviations of Names of Herbaria

B   — Botanisches Museum, Berlin—Dahlem
BM   — British Museum of Natural History, London
E   — Royal Botanic Garden, Edinburgh
G   — Conservatoire et Jardin botaniques, Geneva
K   — The Herbarium, Royal Botanic Gardens, Kew, Surrey,
          London
KW   — Institute of Botany of the Academy of Sciences of the
          Ukrainian SSR, Kiev
L   — Rijksherbarium, Leiden
Linn.   — The Linnean Society of London, London
M   — Botanische Staatsammlung, München
MW   — Herbarium of the Moscow State University, Moscow
P   — Museum National d'Histoire Naturelle, Laboratoire de
          Phanerogamie, Paris
PR   — Botanical Department of the National Museum, Prague
S   — Naturhistoriska Riksmuseum, Botanical Department,
          Stockholm
TAK   — Herbarium of the Tashkent State University, Tashkent
TI   — Botanical Institute, Faculty of Science, University of
          Tokyo, Hongo, Tokyo

TK — Krylov Herbarium of the State University, Tomsk
UPS — Institute of Systematic Botany, University of Uppsala, Uppsala
W — Naturhistorisches Museum, Wien (Vienna)
WRSL — Botanical Institute of the University of Wroclaw, Wroclaw

Family 24. LILIACEAE Juss.

1. Bushy lianes with coriaceous cordate leaves with reticulate venation. Flowers in axillary umbels; fruit—berry .... 19. **Smilax** L. (*S. stans* Maxim.).
+ Herbaceous plants; leaves with longitudinal, parallel or arcuate veins or replaced by acicular cladodes—green branches .... 2.
2. Plants rhizomatous .... 3.
+ Plants bulbous .... 13.
3. Leaves not developed, in form of white scales, and functionally replaced by green acicular branches, borne in clusters in their axils. Rhizome underground, stout, branched. Fruit—berry .... 13. **Asparagus** L.
+ Leaves green, normally developed .... 4.
4. All leaves radical, linear. Flowers in terminal raceme. Fruit—capsule. Rhizome short .... 5.
+ All leaves cauline, broad, elliptical to cordate .... 7.
5. Flowers large, funnel-shaped, with tubular base, yellow or orange, in short few-flowered inflorescence .... 4. **Hemerocallis** L.
+ Flowers relatively small, campanulate, with lobes separated down to base, white, yellowish or dull purple, in long many-flowered raceme .... 6.
6. Flowers in slender lax raceme, white; scape with few bract-like leaves. Roots slender, funiform .... 3. **Anemarrhena** Bge. (*A. asphodeloides* Bge.).
+ Flowers in compact cylindrical raceme; scape entirely leafless. Roots fusiform, thickened .... 2. **Eremurus** M.B.
7. Flowers axillary, pendent. Fruit—berry. Leaves in whorls or alternate, sessile. Rhizome thick, more or less long, horizontal, or shortened .... 8.
+ Flowers in terminal raceme, rarely solitary, terminal. Leaves alternate, rarely clustered in single whorl and then petiolate. Rhizome slender, creeping or thick, vertical .... 9.
8. Flowers solitary or few, tubular; perianth gamophyllous. Leaves in whorls or alternate, or partly alternate and partly in whorls, sessile. Rhizome horizontal .... 17. **Polygonatum** Adans.

+ Flowers solitary, on long filiform pedicel appressed to lower surface of subtending leaf, with open campanulate perianth with free lobes. Leaves alternate, with amplexicaul base. Rhizome shortened with cluster of slender funiform roots ....16. **Streptopus** Michx. [*S. amplexifolius* (L.) DC.].

9. Leaves flat. Flowers in simple slender raceme, or flowers solitary. Fruit—berry. Rhizome slender, creeping ....10.

+ Leaves longitudinally folded (fluted), alternate, sessile. Flowers in dense paniculate inflorescence, stellate. Fruit—capsule. Rhizome thick, short, vertical..1. **Veratrum** L.

10. Leaves aggregated in whorl near tip, obovate, 4 (rarely 5). Flower solitary, terminal, with stellate 4-merous distichous apple-green perianth. Berry glaucescent black... **Paris** L.

+ Leaves 2–3, alternate. Flowers in slender raceme; perianth corolliform, white. Berry red...11.

11. Leaves cordate, petiolate, 2. Flowers very small, 4-merous, stellate .... 15. **Majanthemum** Wigg. [*M. bifolium* (L.) F.W. Schmidt].

+ Leaves lanceolate or elliptical, sessile, 2–3.................................12.

12. Flowers broadly cup-shaped, with gamophyllous 6-toothed perianth, pendent. Leaves with long narrow sheaths, forming tall tube closely encircling slender scape.. ....18. **Convallaria** L. (*C. keiskei* Miq.).

+ Flowers widely patent, 6-merous, stellate, not pendent. Leaves semiamplexicaul............. 14. **Smilacina** Desf. [*S. trifolia* (L.) Desf.].

13. Flowers (1–3) together with leaves enclosed in tubular aphyllous sheath; perianth gamophyllous, with long tube and yellow funnel-shaped 6-partite limb. Capsule ovate; carpels acuminulate and revolute .... **Colchicum** L. (*C. luteum* Baker).

+ Flowers and leaves on exposed stem (not enclosed in aphyllous sheath); perianth free down to base or almost down to base. Capsule undivided at tip ....14.

14. Flowers quite small, aggregate in umbellate inflorescence; before anthesis, enclosed in 2 spathaceous bracts persistent at its base; flowers rarely solitary ....15.

+ Flowers generally large, solitary or clustered in terminal raceme ....16.

15. Flowers yellow, lustrous inside, greenish outside or with green band on back, in few-flowered racemous-umbellate inflorescence; covered with 2 narrow bracts before anthesis .... 5. **Gagea** Salisb.

+ Flowers pink, purple, blue, white or yellowish, in many-flowered capitate umbel; before anthesis enclosed in undivided membranous spathe (hood) .... 6. **Allium** L.

16. Flowers solitary, pendent; perianth lobes oblong, united at base into short tube and abruptly recurved above towards pedicel, divaricate stellately. Stem midlength with 2 petiolate ovate-lanceolate leaves ....
.... 11. **Erythronium** L. [*E. sibiricum* (Fisch. et Mey.) Kryl.].
+ Flowers with free individually caducous segments, funnel- or cup-shaped. Leaves sessile .... 17.
17. Flowers solitary or in few-flowered raceme, with erect lobes, cup-shaped or broadly funnel-shaped but then white and relatively small; anthers basifixed .... 18.
+ Flowers in raceme, more rarely solitary, funnel-shaped, often with lobes greatly recurved; anthers versatile .... 7. **Lilium** L.
18. Flowers broadly cup-shaped, large .... 19.
+ Flowers broadly funnel-shaped, small, white, solitary, rarely 2–8. Leaves linear or linear-filiform, herbaceous ... 12. **Lloydia** Reichb.
19. Flowers erect, solitary; perianth lobes without nectary at base. Bulbs covered with coriaceous scales .... 10. **Tulipa** L.
+ Flowers pendent, solitary or in raceme; perianth lobes with nectary at base and in chequered pattern. Bulbs not covered with coriaceous scales .... 20.
20. Nectary bulging outward, calcariform, on single external perianth lobe larger than rest and recurved; as a result, flowers zygomorphic. Capsule globose, wingless. Flowers many, in raceme. Leaves alternate, amplexicaul ....
.... 8. **Rhinopetalum** Fisch. (*Rh. karelinii* Fisch.).
+ Nectary flat, flowers actinomorphic. Capsule turbinate, winged. Flowers generally solitary, rarely in pairs. Leaves opposite, whorled or alternate .... 9. **Fritillaria** L.

1. **Veratrum** L.
Sp. pl. (1753) 1044.

1. Flowers cervine, pedicel as long as perianth or longer. Leaves glabrous .... 1. **V. nigrum** L.
+ Flowers yellowish-green, pedicel many times shorter than perianth. Leaves puberulent beneath .... **V. lobelianum** Bernh.

1. **V. nigrum** L. Sp. pl. (1753) 1044; Franch. Pl. David. 1 (1884) 910; Forbes and Hemsley, Index Fl. Sin. 3 (1903) 147; Kryl. Fl. Zap. Sib. 3 (1929) 586; Kuzeneva in Fl. SSSR, 4 (1935) 11; Kitag. Lin. Fl. Mansh. (1939) 145; Grub. Konsp. fl. MNR (1955) 91; Fl. Kazakhst. 2 (1958) 105; Pazij in Opred. rast. Sr. Azii [Key to Plants of Mid. Asia], 2 (1971) 11. —**Ic.:** Bot. Mag. tab. 963; Reichb. Ic. Fl. Germ. 10, tab. 423; Fl. Kazakhst. 2, Plate IX, fig. 2.

Described from Europe. Type in London (Linn.).

Forest meadows and fringes, scrub thickets and meadow-covered slopes of mud cones.

IA. **Mongolia:** *East. Mong.* (left bank of Khalkhin-Gol near Derkhin-Tsagan-obo knoll, nor. slope of mud cone 913, meadow steppe, Aug. 16, 1970—Grub. et al).

IIA. **Junggar:** *Cis-Alt.* (right bank of Kairta river, Kuidyn valley 25–30 km north of Kok-Togai settlement, in lower forest belt, July 15, 1959—Yun. and I.-F. Yuan'); *Tarb.* (Dachen town environs [Chuguchak], 1770 m, No. 1628, Aug. 13, 1959—Kuan).

General distribution: Fore Balkh., Jung.-Tarb.; Europe (Cen. and East.), Balk.; West. Sib. (south.), Far East (south.), Nor. Mong. (Mong.-Daur., Fore Hing.), China (Dunbei, Nor., Cent., South-west), Japan.

V. lobelianum Bernh. in Schrad. Neues Journ. Bot. (Göttingen) 2, 2–3 (1808) 356; Kryl. Fl. Zap. Sib. 3 (1929) 587; Kuzeneva in Fl. SSSR, 4 (1935) 13; Fl. Kirgiz. 3 (1951) 24; Grub. Konsp. fl. MNR (1955) 91; Fl. Kazakhst. 2 (1958) 106; Pazij in Opred. rast. Sr. Azii [Key to Plants of Mid. Asia], 2 (1971) 11. —Ic.: Reichb. Ic. Fl. Germ. 10, tab. 938; Fl. Kazakhst. 2, Plate IX, fig. 1.

Described from Europe. Type in Vienna (?).

Larch and birch-larch forests and their fringes, forest meadows.

Found in adjoining regions of Nor. Mongolia and Junggar.

General distribution: Fore Balkh., Jung.-Tarb., Nor. Tien Shan; Europe (Cen. and East.), Mediterr., Balk., Caucasus, West. Sib., East. Sib. (south.), Far East (west.), Nor. Mong. (Fore Hubs., Hang.).

Colchicum L.
Sp. pl. (1753) 341.

C. luteum Baker in Gard. Chron. new ser. 2 (July 1874) 34; Tsernyak. in Fl. SSSR, 4 (1935) 27. —Ic.: Bot. Mag. tab. 6153.

Described from Kashmir. Type in London (K).

Erroneously shown in "Fl. SSSR" (l.c.) as described from Jarkend (Kashgar). In fact, this species was described from the herbarium specimens of T. Thomson from Kashmir preserved in the Herbarium of Kew Gardens and has not been reported from Kashgar. Following Thomson, it was collected from the same site in Kashmir by G. Henderson during his Jarkend expedition, as stated by Baker in the first description, which was probably the source of this error.

2. Eremurus M.B.
Fl. taur.-cauc. 3 (1819) 169.

1. Perianth lobes with 3 nerves; flowers narrowly campanulate, yellowish or purple. Leaves lanceolate-linear, 6–20 mm broad. Capsule about 10 mm in diam 2

+ Perianth lobes invariably with single nerve; flowers broadly campanulate, white or pink, large. Leaves linear, 4–6 mm broad, stem 30–50 cm tall. Capsule 15–20 mm in diam.; seed unequally winged .... 2. **E. anisopterus** (Kar. et Kir.) Rgl.
2. Flowers yellowish, pedicel jointed in upper part. Capsule nigrescent; seed wingless. Stem slender ....
.... 1. **E. altaicus** (Pall.) Stev.
+ Flowers dull purple, pedicel not jointed. Capsule not nigrescent before maturity; seed broadly winged. Stem thick, fleshy ....
.... 3. **E. inderiensis** (M.B.). Rgl.

1. **E. altaicus** (Pall.) Stev. in Bull. Soc. natur. Moscou, 4 (1832) 255; Rgl. in Acta Horti Petrop. 6, 2 (1880) 534; O. Fedtsch. Eremurus (1909) 44; Kryl. Fl. Zap. Sib. 3 (1929) 589; B. Fedtsch. in Fl. SSSR, 4 (1935) 41; Fl. Kazakhst. 2 (1958) 112; Vved. and Kovalevsk. in Opred. rast. Sr. Azii [Key to Plants of Mid. Asia], 2 (171) 20. —*Asphodelus altaicus* Pall. in Acta Ac. Sci. Petrop. 1779, 2 (1783) 258. —**Ic.:** Pall. l.c. tab. 19; O. Fedtsch. l.c. tab. 3.

Described from Altay. Type in Berlin (?).

Exposed rubble and stony mountain slopes in steppe belt and lower part of forest belt, 500–2700 m alt.

IIA. **Junggar:** *Cis-Alt.* (Kandagatai river, among stones, Sept. 14, 1876—Pot.), *Tarb.* (5 km east of Sulugui [south-west of Ulyungur], 530 m, No. 10300, May 21, 1959—A.R. Lee 1959), *Jung.-Alatau* (Albakzyn mountains, Toli district, on slope, No. 2609, 2724, Aug. 7, 1957—Kuan; Ven'tsyuan' [Arasan], south. slope, 2450 m, No. 1454, Aug. 14; same locality, 2200 m, on rubble slope, No. 3471, Aug. 14, 1957—Kuan); *Tien Shan* (Dzhagastai, 1500–2100 m, Aug. 7 and 8, 1877—A. Reg.; midcourse of Khorgos, 1200–1500 m, May 15; Khanakhai mountain range, 1500–2100 m, June 16; Kyzemchek mountains, 2400–2700 m, Aug. 29—1878, A. Reg.; Bainamun near Dzhin, 1500–1800 m, June 5; Tsagan-Usu in Iren-Khabirga mountain range, 1800–2400 m, June 16; Borgaty pass, 2400–2700 m, June 7; Borgaty, 1800 m, July 4—1879, A. Reg.; nor. slope between Barkul' town and Khami town, in larch grove, June 3, 1879—Przew.; near Muzart, 1886—Krasnov; near Santash pass and along Tyubu river, 2100–2450 m, June 2, 1893—Rob.; Kizyl-Zangi village south of Shichan, No. 462, July 23; 6 km south-west of Shichan, 1700 m, exposed slope, No. 2297, July 26—1957, Kuan; 15 km north of Ulistai, No. 1233, Aug. 28, 1957—Kuan; nor. slope of Bogdo-Ul along road from Fukan to sacred lake, lower part of forest belt, along south. rocky slope, April 26, 1959—Yun. and I.-F. Yuan').

General distribution: Jung.-Tarb., Nor. Tien Shan; West. Sib. (south. Irt., Altay).

2. **E. anisopterus** (Kar. et Kir.) Rgl. in Acta Horti Petrop. 2, 2 (1873) 429; ibid. 6, 2 (1880) 534; B. Fedtsch. in Fl. SSSR, 4 (1935) 50, p.p.; Fl. Kazakhst. 2 (1958) 117; Vved. and Kovalevsk. in Opred. rast. Sr. Azii [Key to Plants of Mid. Asia], 2 (1971) 25. —*Henningia anisoptera* Kar. et Kir. in Bull. Soc. natur. Moscou, 15 (1842) 517. —**Ic.:** O. Fedtsch. Eremurus, tab. 17.

Described from Kazakhstan (Fore Balkhash). Type in Leningrad.

Sandy plains and dunes.

IIA. **Junggar:** *Jung. Gobi* (cen.: 10 km nor. of Savan settlement, sand-dunes, No. 3799, Oct. 7, 1956—Ching; Savan district, 5 km south of Paotai, sand-dunes, No. 50, June 10; along road from Savan settlement to Paotai, sand, No. 738, June 10, 1957—Kuan; left bank of Manas river 20 km nor.-west of Savan settlement along road to Paotai state farm, sand ridge, 1957—Yun. et al.); *Dzhark.* (in Ili river valley around Suidun, May 8, 1878—A. Reg.).

General distribution: Aralo-Casp. (south-east.), Fore Balkh.

3. **E. inderiensis** (M.B.) Rgl. in Acta Horti Petrop. 2, 2 (1873) 427; ibid. 6, 2 (1880) 534; O. Fedtsch. Eremurus (1909) 61; Kryl. Fl. Zap. Sib. 3 (1929) 590; B. Fedtsch. in Fl. SSSR, 4 (1935) 43; Fl. Kazakhst. 2 (1958) 114; Fl. Tadzh. 2 (1963) 190; Vved. and Kovalevsk. in Opred. rast. Sr. Azii [Key to Plants of Mid. Asia], 2 (1971) 17. —*E. spectabilis* f. *inderiensis* M.B. Fl. taur.-cauc. 3 (1819) 270 in adnot. —**Ic.:** O. Fedtsch. l.c. tab. 6.

Described from Kazakhstan (West., Indersk hills along Ural river). Type in Leningrad.

Sandy deserts.

IIA. **Junggar:** *Jung. Gobi* (Cen.: between Guchen and Baidaotsyao area, sand, Sept. 18 (30), 1875—Pias.; on way from Savan to Paotai, on sand, No. 737, June 10, 1957—Kuan; left bank of Urungu river 25 km east of Din'syan settlement along road to Ertai, on thin sand terrace covered with winterfat, July 13, 1959—Yun. and I.-F. Yuan'; west.: Sairam-Nor, July 1978—A. Reg.); *Zaisan* (Ch. Irtysh river, left bank, west of Cherek-Tas hill, hummocky sand, June 10, 1914—Schischk.); *Dzhark.* (in Ili river valley near Suidun, May 1878—A. Reg.).

General distribution: Aralo-Casp.; Fore Balkh.; Fore Asia, Mid. Asia (west. Kopet-Dag, south. Pam.-Alay).

3. Anemarrhena Bge.
Enum. pl. China bor. (1832) 66.

1. **A. asphodeloides** Bge. l.c.; Franch. Pl. David. 1 (1884) 303; Forbes and Hemsley, Index Fl. Sin. 3 (1903) 118; Kitag. Lin. Fl. Mansh. (1939), 133; Grub. Konsp. fl. MNR (1955) 91.

Described from North China. Type in Paris. Isotype in Leningrad.

Meadows and forb steppes.

IA. **Mongolia:** *East. Mong.* (Sommet des montagnes, au nord de Sartchy, No. 2910, 1866—David; Khaligakha area, on steppe with sandy soil, June 23; Ul'gen-Gol river, July 7—1899, Pot. and Sold.; 85 km north-east of Erdene-Tsagan [Yugodzyr], forbs-common cattail steppe, July 1, 1971—B. Dashnyam, Z. Karamysheva, et al.); *Ordos* (5 km north-west of Ushin town, meadow, Aug. 3; 30 km south of Dalat town, meadow in Huang He valley around Chzhandanchzhao village, Aug. 10—1957, Petr.).

General distribution: China (Dunbei, North, North-west: Shenxi).

4. Hemerocallis L.
Sp. pl. (1753) 324.

1. Inflorescence branched; flowers 3–8; leafy bracts herbaceous lanceolate- or linear-subulate; perianth lemon-yellow, with 2–3 cm

long tube; inner lobes with unanastomosed nerves. Stem 40–70 cm tall. Leaves 5–15 mm broad. Rhizome often with offshoots (especially in cultivated plants) .... 1. **H. lilio-asphodelus** L.

+ Inflorescence generally simple, 2–5 flowered; flowers often solitary; leafy bracts scarious, ovate; perianth yolk-yellow, darker coloured outside, with 1.5–2 cm long tube; inner lobes with nerves interconnected with transverse anastomoses. Stem 25–50 cm tall. Leaves 3–8 mm broad. Rhizome without offshoots ....
.... 2. **H. minor** Mill.

1. **H. lilio-asphodelus** L. Sp. pl. (1753) 324 excl. var. *β. fulva* L.—*H. flava* L. Sp. pl. ed. 2 (1762) 462; Forbes and Hemsley, Index Fl. Sin. 3 (1903) 115; Kryl. Fl. Zap. Sib. 3 (1929) 591; Stout, Daylilies (1934) 17; Tschernyak. in Fl. SSSR, 4 (1935) 59; Kitag. Lin. Fl. Mansh. (1939) 136; Grub. Konsp. fl. MNR (1955) 92; Fl. Kazakhst. 2 (1958) 117. —**Ic.:** Bot. Mag. tab. 19; Reichb. Ic. Fl. Germ. tab. 1112; Fl. SSSR, 4. Plate V, fig. 4.

Described from Europe. Type in London (Linn.).

Floodplain and forest meadows, meadow slopes and meadow steppes.

IA. Mongolia: *East. Mong.* (between Lykse lake and Ulugui river, Aug. 5; Kulun-Buirnor plain, Lodzin sentry post, on clayey-sandy arid soil, July 9—1899, Pot. and Sold.; 25 km from former Bayan-Tsagan somon along road to Abdrant-Gol, forb meadow steppe with almond grove, Aug. 20, 1970—Grub. et al.).

General distribution: Europe, Mediterranean, West Sib. (south. and Altay west.), Far East (south.), China (Dunbei, Nor., Cent.), Korean peninsula.

2. **H. minor** Mill. Gard. Dict. ed. 8 (1768) No. 2: Franch. Pl. David. 1 (1884) 303; Forbes and Hemsley, Index Fl. Sin. 3 (1903) 116; Stout, Daylilies (1934) 18; Tschernyak. in Fl. SSSR, 4 (1935) 60; Kitag. Lin. Fl. Mansh. (1939) 136; Grub. Konsp. fl. MNR (1955) 92. —*H. graminea* Andr. Bot. Repos. 4 (1802) tab. 244; Maxim. Prim. Fl. Amur. (1859) 478. —*H. flava* L. var. *minor* (Mill.) Hotta in Acta phytotax. geobot. 22 (1966) 40. —*H. flava* auct. non L.: Danguy in Bull. Mus. nat. hist. natur. 20 (1914) 138.

Described from cultivated specimens of Siberian origin. Type in London (BM).

Forb-grass and sandy steppes, steppe and forest meadows, meadow slopes.

IA. Mongolia: *East. Mong.* (Inter Kulussutajewsk et Dolon-Nor, 1870—Lom.; Muni-Ula, forest meadows, common, July 8, 1871—Przew.; Khaligakha area, sandy steppe, June 23; Bilyutai hill, watershed upland, July 4—1899, Pot. and Sold.; Khailar, July 6, 1901—Lipsky; vicinity Manchuria railway station, 1915—Nechaeva; Khailar, true and meadow steppe, 1959—Ivan.; [Tamtsagskii ridge] 28–30 km south-east of Bayan-Burid somon, forb-grass steppe, Aug. 5; 30 km east-south-east of Bayan-Tsagan somon, feather-grass-forb steppe, Aug. 6—1949, Yun.; Dege-Gol and Numurgiin-Gol interfluve region along road southward from Khamar-Daba, forb-grass steppe, Aug. 16, 1970—Grub. et al.).

General distribution: East. Sib. (south.), Far East (south.), Nor. Mong., China (Dunbei, North, North-west, Cent.), Korean peninsula.

Note. Related to preceding species through intermediates; the 2 species are difficult to distinguish in the eastern part of the distribution range of *H. minor*, i.e., Dauria and North and North-east China.

5. Gagea Salisb.

In Koenig and Sims, Ann. Bot. 2, 3 (1806) 555.

1. Bulb covered with fibrous tunic extended above it into sheath, forming more or less long neck surrounding stem base. Leaves alternate. Inflorescence corymbose, paniculate, or flowers solitary ....2.
+ Bulb covered with short, coriaceous or papery, dense tunic, not extended into sheath. Leaves aggregate under inflorescence, rarely alternate. Inflorescence umbellate, rarely corymbose-paniculate ....5.
2. Neck of bulb high, reaching midstem ....3.
+ Neck of bulb low, up to 1 cm tall ....4.
3. Stigma tripartite down to base .... 15. **G. pauciflora** Turcz.
+ Stigma undivided, 3-lobed .... 7. **G. albertii** Rgl.
4. Upper cauline leaves with smaller bulbs in axils; pedicels villous-ciliate .... 10. **G. jaeschkei** Pascher.
+ Cauline leaves without smaller bulbs in axils; pedicels glabrous .... 14. **G. vvedenskyi** Grossh.
5. Cauline leaves alternate and more or less evenly distributed on stem. Inflorescence corymbose-paniculate ....6.
+ Cauline leaves aggregate in upper part of stem under umbellate inflorescence..8.
6. Smaller bulbs foveolate-alveolate. Flowers almost white; ovary stipitate, pyriform ....11. **G. ova** Stapf.
+ Smaller bulbs glabrous. Flowers apple-green; ovary sessile, ovate ....7.
7. Smaller bulbs in axils of cauline leaves. Whole plant with dense puberulent pubescence ....8. **G. bulbifera** (Pall.) Salisb.
+ Axillary small bulbs absent. Whole plant glabrous. Pedicels piliform, rigescent in fruits, erect .... 13. **G. tenera** Pascher.
8. Perianth lobes narrowly acuminate. Seed flat. Radical leaf 5-angled in section, glaucescent ....9.
+ Perianth lobes obtuse or obliquely emarginate; if acute, capped. Seed angular. Radical leaf flat, rarely semicylindrical, fistular ....10.

9. Pedicels arcuately deflexed after anthesis and in fruit. Whole plant glabrous .... 9. **G. divaricata** Rgl.

+ Pedicels invariably erect. Whole plant, including perianth lobes, with fine pubescence outside .... 12. **G. pseudoreticulata** Vved.

10. Inflorescence base with 2 opposite, strongly unequal leaves; rest small, bract-like, pedicels present .... 11.

+ Inflorescence base with only 1 leaf; rest assidenous above, small, bract-like, or altogether absent .... 12.

11. Radical leaf fistular, 3-angled, semicylindrical. Stem terete. Bulbs 2; radical leaf emerging from larger bulb and stem from other much smaller bulb. Pedicels erect in fruit; perianth lobes emarginate at tip, or obliquely abscised, or obtuse .... 1. **G. emarginata** Kar. et Kir.

+ Radical leaf flat, linear, grooved above, distinctly keeled beneath. Stem tetrahedral. Bulb 1. Pedicels deflexed after anthesis; perianth lobes symmetrical, obtuse .... 6. **G. turkestanica** Pascher.

12. Perianth lobes with saccate nectaries at base .... 5. **G. sacculifera** Rgl.

+ Perianth lobes without saccate nectary .... 13.

13. Radical leaf flat, lanceolate or linear-lanceolate, 4–10 mm broad, rarely narrower, 2–3 mm. Numerous fine bulbs around large bulb ....14.

+ Radical leaf narrowly linear or filiform, 1–3 mm, rarely up to 5 mm broad. Bulbs under tunic only 2, strongly unequal, sometimes only 1. Cauline leaves and pedicel crispate-hairy; pedicel in fruit, long, strongly divaricate .... 2. **G. filiformis** (Ledeb.) Kunth.

14. Cauline leaves and pedicels glabrous. Bulb tunic grey or brown; smaller bulbs white .... 15.

+ Cauline leaves and pedicel crispate-hairy. Bulb tunic coriaceous, dark brown; smaller bulbs black .... 4. **G. pseudoerubescens** Pascher.

15. Stem and inflorescence branches flexuous. Smaller bulbs clustered alongside large bulb. Perianth 5–7 mm long .... **G. dschungarica** Rgl.

+ Stem and inflorescence branches erect. Perianth 7–18 mm long .. ..16.

16. Smaller bulbs surround base of large bulb like corona. Perianth lobes yellow with green or brown bands outside .... 3. **G. granulosa** Turcz.

+ Smaller bulbs sessile in clusters alongside large bulb. Perianth lobes green outside .... **G. hiensis** Pascher.

12

Subgenus 1. GAGEA

Section 1. Gagea

G. dschungarica Rgl. in Acta Horti Petrop. 6, 2 (1880) 513; Grossh. in Fl. SSSR, 4 (1935) 75; Fl. Kirgiz. 3 (1951) 41; Fl. Kazakhst. 2 (1958) 121; Uphof in Pl. Life, 15 (1959) 161; Fl. Tadzh. 2 (1963) 216; Vved. in Opred. rast. Sr. Azii [Key to Plants of Mid. Asia], 2 (1971) 30. —Ic.: Fl. SSSR, 4, Plate VI, fig. 2.

Described from East. Kazakhstan (Jung. Alatau). Type in Leningrad.
Slopes along borders of neve basins in middle and upper hill belts.

Occurrence possible in IIA. Junggar since it is distributed in adjoining regions of the USSR, i.e., Jung. Alatau.

General distribution: Jung.-Tarb., Mid. Asia (hilly).

1. **G. emarginata** Kar. et Kir. in Bull. Soc. natur. Moscou, 14 (1841) 851; Grossh. in Fl. SSSR, 4 (1935) 88; Fl. Kirgiz. 3 (1951) 45; Fl. Kazakhst. 2 (1958) 126; Fl. Tadzh. 2 (1963) 219; Vved. in Opred. rast. Sr. Azii [Key to Plants of Mid. Asia], 2 (1971) 32. —*G. arvensis* var. *intermedia* Rgl. in Acta Horti Petrop. 6, 2 (1880) 308. —Ic.: Fl. Tadzh. 2, Plate 45, fig. 6.

Described from East. Kazakhstan. Type in Leningrad.
Steppe and meadow slopes of mountains up to alpine belt.

IIA. Junggar: *Tien Shan* (Almaty gorge up to Kokkamyr, 1200–2100 m, April 28; Kokkamyr upland, upper Sary-Bulak, 1800–2100 m, April 29—1878, A. Reg.; Chungur-Daban pass, 3050 m [Iren-Khabirga mountain range, June 13], 1879—A. Reg.

General distribution: Jung.-Tarb., Nor. and Cent. Tien Shan; Mid. Asia (Pamiro-Alay), West. Sib. (south. Altay).

2. **G. filiformis** (Ledeb.) Kunth, Enum. pl. 4 (1843) 237; Rgl. in Acta Horti Petrop. 6, 2 (1880) 309; Kryl. Fl. Zap. Sib. 3 (1929) 599; Grossh. in Fl. SSSR, 4 (1935) 599; Fl. Kirgiz. 3 (1951) 41; Fl. Kazakhst. 2 (1958) 121; Grub. in Bot. mat. (Leningrad) 19 (1959) 534; Fl. Tadzh. 2 (1963) 217; Vved. in Opred. rast. Sr. Azii [Key to Plants of Mid. Asia], 2 (1971) 31. —*Ornithogalum filiforme* Ledeb. Fl. alt. 2 (1830) 30. —Ic.: Ledeb. Ic. pl. fl. ross. 4, tab. 392.

Described from Altay (west.). Type in Leningrad.
Steppe and meadow slopes from middle to upper mountain belts.

IIA. Junggar: *Tien Shan* (Almaty gorge up to Kokkamyr, 1200–2100 m, April 28; Kokkamyr upland, upper Sary-Bulak, 1800–2100 m, April 29—1878, A. Reg.; Lauzogun, 750 m, April 8; upland in sources of Dzhirgalan and Pilyuchi, 1800 m, April 24—1879, A. Reg.; Kungess Tal, May 1–5, 1908—Merzb.; nor. slope of Bogdo-Ul, Sangunkhe basin, arid Takhuangu gorge, lower part of valley, 1300 m, right bank terrace with clayey soil, April 6, 1954—Mois.; Manas river basin, Danu pass between Ulan-Usu and Danu-Gol rivers, nival belt, under shadowed rocks, July 23, 1957—Yun. et al.); *Jung. Gobi* (east.: Shankhausyan mountain range, in solonetz. moist clay, scattered, June 25, 1879—Przew.).

General distribution: Jung.-Tarb., Nor. Tien Shan; Fore Asia, Mid. Asia (hilly), West. Sib. (south-east., west. Altay).

3. **G. granulosa** Turcz. in Bull. Soc. natur. Moscou, 27 (1854) 112; Kryl. Fl. Zap. Sib. 3 (1929) 596; Grossh. in Fl. SSSR, 4 (1935) 71; Fl. Kazakhst. 2 (1958) 121; Uphof in Pl. Life, 15 (1959) 158. —Ic.: Fl. SSSR, 4, Plate VI, fig. 4, a–c; Fl. Kazakhst. 2, Plate XI, fig. 1.

Described from West. Siberia. Type in Kiev.

Forest meadows and fringes, among scrubs along river-banks.

IIA. Junggar: *Cis-Alt.* (south. slope of Chinese Altay, high upland between Burchum and Kran, Bugotar river, June 11, 1903—Gr.-Grzh.).

General distribution: Arctic (nor.), Europe (East.), West. Sib., East. Sib. (Ang.-Sayan).

G. hiensis Pascher in Lotos, 14 (1904) 124; Kryl. Fl. Zap. Sib. 3 (1929) 600; Grossh. in Fl. SSSR, 4 (1935) 72; Kitag. Lin. Fl. Mansh. (1939) 135; Grub. Konsp. fl. MNR (1955) 92. —Ic.: Fl. SSSR, 4, Plate VI, fig. 6, a–b.

Described from East. Siberia (Dauria). Type in Leningrad.

Reported for Nor. Mongolia (Hent.).

4. **G. pseudoerubescens** Pascher in Feddes repert. 2 (1906) 67; Grossh. in Fl. SSSR, 4 (1935) 75; Fl. Kirgiz. 3 (1951) 41; Fl. Kazakhst. 2 (1958) 121; Grub. in Bot. mat. (Leningrad) 19 (1959) 534; Uphof in Pl. Life, 16 (1960) 166; Fl. Tadzh. 2 (1963) 218; Vved. in Opred. rast. Sr. Azii [Key to Plants of Mid. Asia], 2 (1971) 31. —*G. arvensis α typica* Rgl. and *β pusilla* Rgl. in Acta Horti Petrop. 6, 2 (1880) 308, p.p. —Ic.: Fl Kazakhst. 2, Plate XI, fig. 3.

Described from Mid. Asia (Tien Shan). Type in Leningrad.

Forest and coastal meadows, forest glades and fringes, among scrubs, on meadow slopes.

IIA. Junggar: *Tien Shan* (on nor. slope [near Koshety-Daban pass] in forest, June 11 and 12, 1877—Pot.; Borborogusun picket, 1200 m, March 2; Kokkamyr upland, upper Sary-Bulak, 1800–2100 m, April 29—1878, A. Reg.; upland near sources of Dzhirgalan and Pilyuchi, 1800 m, April 24; Nilki brook of Kash river, 2100 m, June 8—1879, A. Reg.; Santai, upper Shigou, terrace above floodplain, 2000 m, May 27, 1952—Mois.; Bogdo-Ula, near sacred lake south of Fukan, grassy spruce forests, in glades, April 26, 1959—Yun.).

General distribution: Jung.-Tarb., Nor. Tien Shan; Mid. Asia, (West. Tien Shan, Pamiro-Alay).

5. **G. sacculifera** Rgl. in Acta Horti Petrop. 6, 2 (1880) 510; Grossh. in Fl. SSSR, 4 (1935) 78; Fl. Kazakhst. 2 (1958) 122. —Ic.: Fl. SSSR, 4, Plate VI, fig. 1.

Described from Junggar (Tien Shan). Type in Leningrad.

Steppe mountain slopes.

IIA. Junggar: *Tien Shan* (Iren-Khabirga mountain range, south-east of Shikho, 900–1500 m, March 25, 1878—A. Reg., typus!; Sarybulak, 1200–1800 m, April 23, 1878—A. Reg.).

General distribution: Jung. Alatau.

6. **G. turkestanica** Pascher in Bull. Soc. natur. Moscou, n.s. 19 (1907) 365; Fl. Kirgiz. 3 (1951) 42; Uphof in Pl. Life, 16 (1960) 164; Vved. in Opred. rast. Sr. Azii [Key to Plants of Mid. Asia], 2 (1971) 32.—*G. capusii* auct. non Terr.: Grossh. in Fl. SSSR, 4 (1935) 81, p.p.; Fl. Kazakhst. 2 (1958) 123. —*G. parva* Vved. ex Grossh. in Fl. SSSR, 4 (1935) 82, descr. ross.; Fl. Kazakhst. 2 (1958) 124; Fl. Tadzh. 2 (1963) 220. —Ic.: Fl. Kazakhst. 2, Plate XI, Figs. 5 and 6.

Described from Mid. Asia (West. Tien Shan). Type in Leningrad.
Loessial and clayey slopes.

IIA. Junggar: Tien Shan (Bagaduslun, east. tributary of Dzhin river, 2100–2750 m, June 4, 1879—A. Reg.).

General distribution: Fore Balkh. (Chu-Iliisk.), Jung.-Tarb., Nor. Tien Shan; Mid. Asia (West. Tien Shan).

Subgenus 2. HORNUNGIA (Bernh.) Pascher

Section 1. Platyspermum Boiss.

7. **G. albertii** Rgl. in Acta Horti Petrop. 6, 2 (1880) 512; Grossh. in Fl. SSSR, 4 (1935) 103; Fl. Kirgiz. 3 (1951) 47; Fl. Kazakhst. 2 (1958) 130; Uphof in Pl. Life, 16 (1960) 171; Vved. in Opred. rast. Sr. Azii [Key to Plants of Mid. Asia], 2 (1971) 35. —Ic.: Fl. Kazakhst. 2, Plate XI, fig. 12.

Described from Junggar (Tien Shan). Type in Leningrad.
Steppe and meadow slopes, rock crevices, from desert to alpine belt.

IIA. Junggar: *Tien Shan* (Karkarausu south-east of Shikho, 300 m, March 27, 1878—A. Reg., typus!; Kul'dzha, April 1879; Bagaduslun, east. tributary of Dzhina, 2100–2750 m, June 4, 1879—A. Reg.); *Jung. Gobi* (nor.: along bank of Urungu river, on rocks, common and abundant, April 24, 1879—Przew.; south.: 5 km nor. of Kuitun, No. 10214, April 23, 1959—A.R. Lee (1959)).

General distribution: Fore Balkh. (Balkh.-Alak.), Jung.-Tarb., Nor. Tien Shan.

8. **G. bulbifera** (Pall.) Salisb. in Koenig and Sims, Ann. Bot. 2, 3 (1806) 557; Rgl. in Acta Horti Petrop. 6, 2 (1880) 511; Kryl. Fl. Zap. Sib. 3 (1929) 600; Grossh. in Fl. SSSR, 4 (1935) 108; Fl. Kirgiz. 3 (1951) 47; Fl. Kazakhst. 2 (1958) 131; Uphof in Pl. Life, 16 (1960) 168; Vved. in Opred. rast. Sr. Azii [Key to Plants of Mid. Asia], 2 (1971) 36. —*Ornithogalum bulbiferum* Pall. Reise, 2 (1773) 736. —Ic.: Fl. SSSR, 4, Plate VIII, fig. 11.

Described from Caspian region (Astrakhan). Type in Leningrad.
Loessial foothills and mountain slopes in semideserts and steppes.

IIA. Junggar: *Tien Shan* (Almaty valley nor.-west of Kul'dzha, 900–1200 m, April 22, 1878—A. Reg.; Kul'dzha, 1906—Muromskii; nor. foothills of Bogdo-Ul 10–15 km nor.-east of Urumchi along road to Fukan, wormwood forest on loess, April 26, 1959—Yun.).

General distribution: Aralo-Casp., Fore Balkh., Jung.-Tarb., Nor. Tien Shan; Europe (south. Europ. USSR), Fore Asia, Caucasus, West. Sib. (south.).

9. **G. divaricata** Rgl. in Acta Horti Petrop. 6, 2 (1880) 510; Grossh. in Fl. SSSR, 4 (1935) 95; Fl. Kazakhst. 2 (1958) 128; Uphof in Pl. Life, 16 (1960) 168; Fl. Tadzh. 2 (1963) 226; Vved. in Opred. rast. Sr. Azii [Key to Plants of Mid. Asia], 2 (1971) 34. —Ic.: Fl. Tadzh. 2, Plate 45, figs. 7, 8; Fl. SSSR, 4, Plate VIII, fig. 8.

Described from Junggar (Ili river valley). Type in Leningrad.

Sandy deserts.

IIA. **Junggar:** *Dzhark.* (Suidun environs, May 7, 1878—A. Reg., syntypus!).
General distribution: Fore Balkh. (Ili river sand); Mid. Asia (plains).

10. **G. jaeschkei** Pascher in Lotos, 24 (1904) 128; Fl. Kazakhst. 2 (1958) 132; Uphof in Pl. Life, 16 (1960) 171; Ikonnikov, Opred. rast. Pamira [Key to Plants of Pamir] (1963) 85; Fl. Tadzh. 2 (1963) 228; Vved. in Opred. rast. Sr. Azii [Key to Plants of Mid. Asia], 2 (1971) 36. —*G. pamirica* Grossh. in Fl. SSSR, 4 (1935) 108, 738; Uphof in Pl. Life, 16 (1960) 168. —Ic.: Fl. SSSR, 4, Plate VI, fig. 18 (sub *G. pamirica*); Fl. Kazakhst. 2, Plate XI, fig. 14.

Described from Himalayas (nor.-east. India). Type in Berlin (?).

Grasslands, rocky cornices, rubble and clayey slopes in alpine belt.

IIIC. **Pamir:** below Ucha pass, along clayey descent, June 17, 1909—Divn.).
General distribution: Jung. Alatau, Nor. and Cent. Tien Shan, East. Pam.; Middle Asia (West. Tien Shan, east. Pamiro-Alay).

11. **G. ova** Stapf in Denkschr. Ak. Wiss. (math.-naturw., Wien) 50, 2 (1885) 16; Grossh. in Fl. SSSR, 4 (1935) 109; Fl. Kirgiz. 3 (1951) 48; Fl. Kazakhst. 2 (1958) 133; Fl. Tadzh. 2 (1963) 232; Vved. in Opred. rast. Sr. Azii [Key to Plants of Mid. Asia], 2 (1971) 37. —Ic.: Fl. Tadzh. 2, Plate 45, figs. 1–2.

Described from Fore Asia (Iran). Type in Vienna.

Clayey and loessial slopes and trails in desert and steppe mountain belts.

IIA. **Junggar:** *Tien Shan* (Borborogusun, 900–1200 m, April 27, 1879—A. Reg.
General distribution: Aralo-Casp., Fore Balkh., Nor. Tien Shan; Fore Asia, Mid. Asia.

12. **G. pseudoreticulata** Vved. in Fl. Turkm. 1 (1932) 268; Grossh. in Fl. SSSR, 4 (1935) 100; Fl. Kazakhst. 2 (1958) 129; Fl. Tadzh. 2 (1963) 227; Vved. in Opred. rast. Sr. Azii [Key to Plants of Mid. Asia], 2 (1971) 33. —Ic.: Fl. SSSR, 4, Plate VIII, fig. 9.

Described from Mid. Asia. Type in Tashkent.

Desert and steppe foothills.

IIA. **Junggar:** *Tien Shan* (ravine mouth of Taldy river, 900–1050 m, May 15, 1879—A. Reg.).
General distribution: Fore Asia, Mid. Asia.

13. **G. tenera** Pascher in Lotos, 24 (1904) 128; Grossh. in Fl. SSSR, 4 (1935) 109; Fl. Kirgiz. 3 (1951) 47; Fl. Kazakhst. 2 (1958) 132; Uphof in Pl. Life, 16 (1960) 174; Fl. Tadzh. 2 (1963) 237; Vved. in Opred. rast. Sr. Azii [Key to Plants of Mid. Asia], 2 (1971) 39. —*G. chlorantha* auct. non Schult.: Rgl. in Acta Horti Petrop. 6, 2 (1880) 511, p.p. —Ic.: Fl. Tadzh. 2, Plate 45, fig. 4.

Described from Mid. Asia. Type in Leningrad.

Loessial foothills.

IIA. **Junggar:** *Dzhark.* (Kul'dzha, April 20, 1877—A. Reg.).
General distribution: Fore Balkh. (Chu-Ili); Jung.-Tarb., Nor. Tien Shan; Fore Asia (north), Caucasus (east. Transcaucasus), Mid. Asia (hilly).

14. **G. vvedenskyi** Grossh. in Fl. SSSR, 4 (1935) 107, 737; Uphof in Pl. Life, 16 (1960) 171; Ikonnik. Opred. rast. Pamira [Key to Plants of Pamir] (1963) 86; Fl. Tadzh. 2 (1963) 228; Vved. in Opred. rast. Sr. Azii [Key to Plants of Mid. Asia], 2 (1971) 36. —*G. korshinskyi* Grossh. in Fl. SSSR, 4 (1935) 77, 735, sub sect. *Gagea!*; Uphof in Pl. Life, 16 (1960) 172. —Ic.: Fl. SSSR, 4, Plate VI, figs. 17, 19.

Described from Mid. Asia. Type in Tashkent (TAK).

On rubble and rocky steppe slopes, steppe meadows; up to alpine belt.

IIA. **Junggar:** *Tien Shan* (Aryslyn, nor. tributary of Kash river, 2700–3050 m, July 12; apex of Taldy river ravine, 2100–2700 m, May 15—1879, A. Reg.).
General distribution: Cent. Tien Shan?, East. Pam.; Mid. Asia (hilly).

Section 2. Plecostigma (Turcz.) Pascher

15. **G. pauciflora** Turcz. in Bull. Soc. natur. Moscou, 1 (1838) 102, nomen; Ledeb. Fl. Ross. 4 (1852) 143; Franch. Pl. David. 1 (1884) 308; Forbes and Hemsley, Index Fl. Sin. 3 (1903) 138; Danguy in Bull. Mus. nat. hist. natur. 20 (1914) 142; Kryl. Fl. Zap. Sib. 3 (1929) 602; Grossh. in Fl. SSSR, 4 (1935) 111; Grub. Konsp. fl. MNR (1955) 92; Uphof in Pl. Life, 16 (1960) 175. —*Plecostigma pauciflorum* Turcz. in Bull. Soc. natur. Moscou, 27, 2 (1854) 113. —*Szechenyia lloydioides* Kanitz, A növénytani...(1891) 60; idem in Széchenyi, Wissensch. Ergebn. 2 (1898) 735. —*Tulipa* sp. aff. *T. eduli* Baker, Hemsley in J. Linn. Soc. (London) Bot. 30 (1894) 139; ej. Fl. Tibet (1902) 199. —Ic.: Kanitz, l.c. (1898) plate. VII, fig. 1, 1–3.

Described from East. Siberia. Type in Leningrad.

Steppe and meadow slopes, grass steppes and arid meadows, often in sandy soil, coastal meadows and rock crevices, up to alpine belt (4300 m).

IA. **Mongolia:** *Mong. Alt.* (Khan-Taishiri mountain range, south. slope in upper Shine-Us brook [nor. of Khalyun somon], 2380–2450 m, in forest on rocks of nor. exposure, June 18, 1971—Grub. et al.); *East. Mong.* (Tsagan-Balgas environs, on mountains, May 3, 1831—Ladyzh.; Muni-Ula, in valleys, April 21, 1872—Przew.; "Vallée du Kéroulen, steppes, May 29, 1896, Chaff."—Danguy, l.c.; between Dabat and Mandybai salt lakes, May 30, 1899—Pot. and Sold.; near Mandybai salt lakes, May 30, 1899—Pot. and Sold.; near Manchuria station, steppe (somewhat rocky) slopes of mounds, May 6, 1908, Kom.; nor.-east. bank of Buir-Nur lake near Khalkhin-Gol estuary, arid meadows on turf-covered dunes, June 9, 1944—Yun.; Khailar town, Sishan' mounds, 580–600 m, meadows, sandy soil, No. 519, July 6, 1951—B. Skvortzov, T.-t. Li, A. Baranov; Khailar, feather-grass—wild rye steppe, 1959—Ivan.; Choibalsan somon, 4 km west of Khabirga railway station, koeleria steppe atop mounds, May 20, 1957—Dashnyam); *Alash.* *Gobi* (Alashan mountain range, Tsuburgan-Gola gorge, April 27, 1908—Czet.).

IIIA. Qinghai: *Nanshan* (North Tetung mountain range, in meadows, rare, June 4, 1872—Przew.; "In m. Nan-san collinis circa Tschakoye, June 10, 1879, Széchenyi"—Kanitz, l.c.; in San'chuan' valley, on valley floor, 1800 m, clayey-sandy soil, April 15; Lanchzha-Lunva river valley near Shënpyn village, May 14—1885, Pot.; Chadzhi gorge, May 12, 1890—Gr.-Grzh.; Humboldt mountain range, nor. slope, Blagodatnyi spring, alpine meadow and rocks, 3050–3650 m, June 3, 1894—Rob.; Burkhan-Budda mountain range, nor. slope, Nomokhun-Gola gorge, 3050–3950 m, along mountain descents on clay, May 18–19, 1900—Lad.); *Amdo* (along Huang He river near Churmyn estuary, 2600–2700 m, in gorge, silty soil, rare, May 19, 1880—Przew.).

IIIB. Tibet: *Weitzan* [left bank of Dychu (in midcourse), 3950 m, on precipices among shrubs, June 24, 1884—Przew.; "Sharakuyi-gol, mountain slope at 4200 m, lat. N 35°50', long. E 93°27', May 29, 1892, Rockhill"—Hemsley, l.c.; Amnen-Kor mountain range, south. slope, 4100–4250 m, along meadow-covered slopes of hills, June 9, 1900—Lad.; Huang He basin, left bank of Sarchu brook, 4100 m, on clayey hill descents and along valleys, May 25, 1901—Lad.].

General distribution: West. Sib. (Altay), East. Sib. (south.), Far East (south.), Nor. Mong. (Hent., Hang., Mong.-Daur.), China (North-West: south-west Gansu).

6. Allium L.[1]

Sp. pl. (1753) 244; ej. Gen. pl. ed. 5 (1754) 143.

1. Leaves with lanceolate, oblong-ovate or oblong-elliptical blades, gradually or abruptly narrowed into petiole, passing into sheath and surrounding scape base. Bulbs elongated cylindro-conical with reticulate tunic. Inflorescence many-flowered, globose or semiglobose .... 2.

+ Leaves with linear, cylindrical (sometimes fistular), semicylindrical or bristle-like convoluted blades, without petiole, passing directly into sheath .... 4.

2. Leaf blades gradually narrowing into petiole; margin smooth; if highly serrulate, flowers pink .... 3.

+ Leaf blades abruptly narrowing into petiole; margin very shortly dentate-ciliate (under binoculars). Flowers pale yellow, or albescent .... 1. **A. ovalifolium** Hand.-Mazz.

3. Leaf blades (2) 3–6 (8) cm broad, with smooth margin. Flowers ivory-white .... **A. victorialis** L.

---

[1]By T.V. Egorova.

18

+ Leaf blades 1.5–2 cm broad, with highly serrulate margin. Flowers pink ....2. **A. prattii** C.H. Wright.

4. Scape robust, thick, succulent, up to 0.7–3 cm in diam. Leaves cylindrical, fistular, (3) 5–20 mm broad. Bulbs usually large, 1.5–5 cm in diam., with brown or cervine coriaceous undivided tunic ....5.

+ Scape slender, rarely thickened and then leaves flat; if cylindrical, leaves narrower ....6.

5. Inflorescence globose, more or less dense; pedicel 2 (3) times longer than perianth, with few bracts at base; perianth stellate, lobes white, opaque; stamens slightly longer than perianth. Bulbs oblong-ovoid, 1.5–2.5 cm in diam. Leaves (3) 5–10 mm broad .... 63. **A. galanthum** Kar. et Kir.

+ Inflorescence very dense, capitate; pedicel shorter than perianth or as long, without bracts at base; perianth campanulate, lobes pale yellow, lustrous; stamens nearly twice or more longer than perianth. Bulbs ovoid or oblong-ovoid, 2–4 cm in diam. Leaves 8–20 mm broad .... 64. **A. altaicum** Pall.

6. Bulbs adnate to more or less well-developed rhizome and clustered on it in small numbers, rarely 2–3, occasionally solitary, generally narrowly oblong-conical or oblong-cylindrical, rarely elongated- and oblong-ovoid or ovoid ....7.

+ Rhizome absent; bulbs solitary, broadly ovate, globose, ovoid, occasionally oblong-ovoid, or (in *A. fasciculatum*—species from South. Tibet) neither bulbs nor rhizome developed but only several thickened fleshy roots seen ....72.

7. Outer tunic of bulb more or less distinctly reticulate-fibrous, with obliquely entangled fibres; network sometimes not visible to naked eye and, moreover, may be somewhat damaged but, in both cases, entire reticulate tunic or at least some persistent parts of it distinct under binoculars; sometimes bulb tunic very greatly damaged and vaguely reticulate. Filaments of inner stamens invariably enlarged ....8.

+ Outer tunic of bulbs undivided, nearly undivided, or laciniated (sometimes very greatly) into parallel fibres. Filaments of inner stamens may/may not be enlarged. ..../..................25.

8. Bulb tunic intensely fibrous, filamentous (not rarely subtomentose), extremely indistinctly reticulate; bulbs about 0.5 cm in diam., nearly invisible, seated on rhizome, form characteristic, very compact mat. Roots numerous, thickened, subfuniform, with dense albescent hairs ....9.

+ Bulb tunic usually distinctly reticulate, not filamentous; bulbs thicker, usually elongated-conical or cylindro-conical, more or less clearly visible, 1–3 or more seated on rhizome. Roots not thickened, not funiform, subglabrous .... 10.

9. Perianth lobes pink or pale pink, oblong-ovate, subobtuse or subacute, 4–5 (6) mm long; filaments of inner stamens enlarged for 1/2 length, toothed .... 13. **A. polyrhizum** Turcz. ex Rgl.

+ Perianth lobes pinkish-violet, oblong-lanceolate, acute or shortly acuminate, 6–7.5 mm long; filaments of inner stamens enlarged for 2/3 or 3/4 length, toothed, sometimes without teeth .... 19. **A. subangulatum** Rgl.

10. Bulb tunic reddish-violet-brown (colour of 'red wood'). Pedicel base without bracts. Perianth lobes pink-purple .... 11.

+ Bulb tunic brown or dark grey. Pedicel base with few or several bracts .... 12.

11. Spathe with long, (0.7) 1–2 cm long, beak. Pedicel usually 2–2.5 times, rarely slightly longer than perianth, very slender; inflorescence somewhat lax .... 14. **A. przewalskianum** Rgl.

+ Spathe with very short, barely visible beak (0.1–0.2 cm, extremely rarely 0.5 cm long). Pedicel as long as perianth or slightly shorter or longer, thickened; inflorescence dense ...... 17. **A. stoliczkii** Rgl.

12. Stamens 1/2 or 1/3 shorter than perianth lobes (nor exserted); filaments of inner stamens without teeth, rarely with poorly developed teeth .... 13.

+ Stamens longer than perianth (exserted); filaments of inner stamens usually with teeth, rarely without teeth .... 16.

13. Pedicels shorter than perianth, equal in size; flowers strongly coloured, violet-blue; perianth lobes 6.5–8 mm long, oblong-elliptical, obtuse; filaments of inner stamens without or with poorly developed teeth .... 8. **A. kansuense** Rgl.

+ Pedicels 1.5–3 times longer than perianth; if equal, pedicels unequal in size; flowers white, pinkish or yellowish; filaments of inner stamens without teeth .... 14.

14. Pedicels unequal (5–12 mm long in solitary inflorescence). Scape foliate for most part. Leaf blades emerging at different levels, convoluted like setae. Flowers narrowly campanulate; perianth lobes 8–9 mm long, albescent or pale pink, with dark purple nerve .... 20. **A. teretifolium** Rgl.

+ Scape foliate in lower part. Leaves clustered at scape base, with leaf blades emerging nearly at same level. Perianth broadly campanulate or substellate .... 15.

15. Bulbs oblong-conical, clustered in small numbers on rhizome; bulb tunic distinctly reticulate. Perianth broadly campanulate, lobes 5–7 mm long, pinkish or ivory-white (at anthesis), with distinct dull purple nerve, elliptical, broadly rounded at tip,

narrowing abruptly into recurved acute beak ....
.... 12. **A. oreoprasum** Schrenk.

+ Bulbs narrowly oblong-conical, generally 1–3 adnate to rhizome; bulb tunic usually not distinctly reticulate (but distinct under binoculars). Perianth substellate; lobes white, usually with faint pinkish shade, greenish nerve, lanceolate or elliptical, obtuse or subacute at tip, without acute recurved tip .... 11. **A. odorum** L.

16 (12). Perianth lobes pale yellow, sometimes with faint pinkish violet shade at tip and centre. Leaf blades flat or cylindrical .... 17.

+ Perianth lobes pink or violet-blue. Leaf blades generally flat...18.

17. Leaves cylindrical, fistular, glabrous, 1–5 mm broad. Bulbs 5–8 mm long. Filaments of inner stamens with long acute teeth on both sides, often as long as perianth lobes ....
.... 9. **A. leucocephalum** Turcz. ex Ledeb.

+ Leaves flat, (1) 2–5 mm broad, margin very finely serrate-ciliate. Bulbs short, 2–4 cm long. Filaments of inner stamens with teeth, usually much shorter than perianth lobes...................................
.... 7. **A. flavidum** Ledeb.

18. Scape foliate at base. Leaves with short sheaths and blades emerging almost at same level. Flowers dark pinkish-violet, sky-blue or blue. Inflorescence semiglobose, lax, relatively few-flowered .... 19.

+ Scape foliate for 1/3 or 1/2 height. Leaves with elongated sheaths and blades emerging at different levels. Flowers pink or pale pink. Inflorescence globose, compact, many-flowered .... 20.

19. Flowers sky-blue or blue; perianth lobes oblong-elliptical, rounded at tip; filaments of inner stamens without or with barely developed teeth. Bulbs narrow, 2–6 mm in diam.; tunic with faintly visible reticulum distinctly visible under binoculars. Leaves linear, 1–2 mm broad ....5. **A. cyaneum** Rgl.

+ Flowers dark pinkish-violet; perianth lobes oblong-elliptical or lanceolate, acute or obtuse at tip, with short recurved cusp; filaments of inner stamens with long narrow teeth. Bulbs up to 10 mm in diam., tunic with distinct slender fleshy reticulum visible even without binoculars. Leaves semicylindrical, about 1 mm broad .... 6. **A. eduardii** Stearn.

20. Leaves semicylindrical, (0.5) 1–2 mm broad, glabrous. Inflorescence globose or subglobose, usually compact; pedicel 1.5–2 (3) times longer than perianth; perianth lobes pink with purple nerve; filaments of inner stamens with acute, erect teeth, occasionally without teeth.......................... 4. **A. clathratum** Ledeb.

+ Leaves flat, (2) 3–6 mm broad .... 21.

21. Inflorescence very dense, globose or subglobose; pedicel short, shorter than perianth or as long, rarely slightly longer, almost invisible under compact flowers. Alpine plant ....22.

+ Inflorescence usually globose, lax, with distinct pedicels; latter 1.5–3 times longer than perianth ....23.

22. Filaments of inner stamens with short teeth on both sides or only on one side; style 5–6 mm long, distinctly exserted from perianth .... 3. **A. amphibolum** Ledeb.

+ Filaments of inner stamens without teeth; style very short, barely exserted from perianth .... 15. **A. schrenkii** Rgl.

23. Perianth lobes pinkish-lilac, with sharp dark purple nerve; filaments of inner stamens with teeth, sometimes bifurcated; breadth of filament at base more than length of part not split into teeth .... 16. **A. splendens** Willd. ex Schult. et Schult. f.

+ Perianth lobes pink with faintly coloured nerve; filaments of inner stamens without teeth or with variously developed teeth but breadth of filaments at base less than length of part not split into teeth ....24.

24. Stamens usually twice longer than perianth; filaments of inner stamens with well-developed, acute, erect teeth. Pedicels generally slender, 0.2–0.3 mm broad; perianth lobes 3.5–4 (5) mm long. Bulbs conical, 3–4 (6) cm long .... 10. **A. lineare** L.

+ Stamens not more than 1.5 times longer than perianth; filaments of inner stamens generally with poorly developed, obtuse teeth, set obliquely erect or laterally, or teeth absent; pedicels usually thickened, about 0.5 mm broad; perianth lobes 4–5.5 mm long. Bulbs subcylindrical, 6–8 (10) mm long .... 18. **A. strictum** Schrad.

25 (7). Stamens 1/4–1/3 or 2–4 times shorter than perianth; filaments greatly enlarged and 1/3 or 1/2 of enlarged upper part connate forming fairly high corona (ring) and adnate to perianth; pedicel base without bracts; perianth campanulate; style with shortly tripartite stigma. Bulbs almost invisible, elongated-cylindrical, with undivided or fibrous outer tunic ....26.

+ Stamens longer or shorter than perianth, may/may not be enlarged, free or somewhat connate and adnate to perianth but corona not formed above connate part, rarely faint corona formed; stigma capitate or flat ....31.

26. Inflorescence clustered, lax, few-flowered; pedicels equal; perianth lobes 5–7 mm long, opaque, pinkish-violet, with sharp, darker coloured nerve, outer ones shorter than inner. Leaves flat or longitudinally folded, not more than 2 mm broad ....61. **A. weschnjakowii** Rgl.

+ Inflorescence oval, ovate, rarely globose, compact; pedicels unequal, 3–4 times shorter in outer than inner flowers; perianth lobes 8–17 mm long, lustrous (at least in lower half), yellow or purple, with faint nerve barely distinguishable from background colour, outer ones longer than inner. Leaves broader .... 27.

27. Leaves flat^2 or longitudinally folded, 5–15 mm broad. Perianth lobes lanceolate or oblong-lanceolate, (12) 15–17 mm long, with greatly elongated yellow acute tip; pink or purple in upper half or throughout at some stage of flowering; stamens 1/2 of perianth, free, unenlarged part very short and anthers almost sessile on anther corona .... 60. **A. semenovii** Rgl.

+ Leaves fistular or cylindrical. Perianth lobes shorter, rounded at tip, subobtuse or subacute, very rarely extended .... 28.

28. Inflorescence dense, subglobose; as a result, pedicels appear equal at first glance, although on closer scrutiny, inner 3–4 times longer than outer ones .... 29.

+ Inflorescence lax, ovate or oval, with distinctly unequal pedicels .... 30.

29. Perianth lobes golden-yellow, sometimes with light pink shade, broadly rounded at tip, inner slightly shorter than outer; stamens 1/4 or 1/3 shorter than perianth lobes; free (not connate) part of filaments about 2 mm long; style 4–5.5 mm long .... 57. **A. chalcophengos** Airy-Shaw.

+ Perianth lobes purple to reddish-pink, subobtuse or subacute at tip, inner considerably shorter than outer; stamens (2) 3–4 times shorter than perianth; free part of filaments less than 1 mm long; style about 1.5 mm long .... 59. **A. monadelphum** Less. ex Kunth.

30. Perianth lobes oblong- or elongated-lanceolate, usually with extended acute or subacute tips; perianth initially yellow, erubescent later, again turning yellow at anthesis .... 58. **A. fedtschenkoanum** Rgl.

+ Perianth lobes elongated-ovate, rounded at tip, not extended; perianth dark purple but turning yellow at anthesis (even in herbarium) .... 56. **A. atrosanguineum** Kar. et Kir.

31 (25). Leaves fistular, cylindrical or semicylindrical; spathe with short (less than 5 mm) beak .... 32.

---

^{2}The nature of the leaf blade is not always easy to establish in herbarium specimens, i.e., whether the leaf is flat, cylindrical or fistular. Sometimes longitudinally folded leaves are treated as fistular or cylindrical. Attention should be paid to the blade margins: in flat leaves, they are fine, usually acute, sometimes serrulate; in fistular and cylindrical leaves (flattened in herbarium specimens), more or less rounded.

+ Leaves flat, longitudinally convoluted bristle-like, filiform or terete[3], sometimes cylindrical or semicylindrical but then spathe with long (1–2 cm) beak .... 38.

32. Stamens longer than perianth, exserted; flowers yellow or pale yellow. Bulbs oblong-cylindrical or oblong-conical, with brown outer and inner tunic. Inflorescence globose, very compact ..... 33.

+ Stamens shorter than perianth; flowers purple- or violet-pink, pink, pale pink, pale yellow or white. Bulbs ovate, oblong-ovate or ovate-conical, with nigrescent outer and nearly white inner tunic .... 34.

33. Bulbs oblong-conical, with lustrous tunic. Leaves 4–7, with elongated sheaths; scape foliate for 1/4 height. Perianth lobes pale yellow, opaque or somewhat lustrous. Spathe without black dots .... 36. **A. condensatum** Turcz.

+ Bulbs oblong-cylindrical, with opaque tunic. Leaves 2–3, with short sheaths, clustered at scape base. Perianth lobes golden-yellow, lustrous. Spathe with numerous black dots ....
.... 34. **A. chrysanthum** Rgl.

34. Inner perianth lobes longer than outer. Flowers pale yellow ....
.... 51. **A. herderianum** Rgl.

+ Perianth lobes equal .... 35.

35. Pedicels unequal: considerably shorter in outer than inner flowers; perianth lobes (7) 10–12 (14) mm long; filaments quite prominently enlarged for 1/3 or 1/2 length .... 36.

+ Pedicels equal or nearly so; perianth lobes 3–7 mm long; filaments somewhat enlarged at base .... 37.

36. Perianth lobes purple- or violet-pink. Scape, sheath and leaf blade glabrous .... 55. **A. schoenoprasum** L.

+ Perianth lobes pink, pale pink or albescent. Scape, leaf sheath and leaf blade (along projecting nerves) scabrous ....
.... 53. **A. karelinii** Poljak.

37. Inflorescence fascicled, lax, few-flowered; pedicels as long as perianth or 2 (3) times longer; perianth lobes pink but later sometimes fade and become albescent, remaining pink only along nerves, 5–7 mm long .... 54. **A. oliganthum** Kar. et Kir.

+ Inflorescence globose, very compact, many-flowered; pedicels shorter than perianth; perianth lobes 3–4 mm long, white ....
.... 52. **A. juldusicolum** Rgl.

38 (31). Bulb tunic (outer as well as inner) brown, rust-coloured, rarely ochreous or pale brown, coriaceous, undivided or sometimes slightly laciniated upward, very rarely divided into broad fibres

---

[3]In terete, unlike semicylindrical leaves, margins of leaf blades thin as in flat leaves.

throughout length. Filaments of inner stamens not enlarged or, rarely, enlarged, but then their enlarged portions connate for half their length and adnate to perianth .... 39.

+   Outer tunic of bulb (often torn while collecting plant specimens) nigrescent, grey or brownish-grey, membranous or compact (but not coriaceous), inner albescent, scarious; bulb tunic undivided or sometimes slightly lacerated or fibrous (in latter case, blackish-brown or light brown). Filaments of inner stamens not enlarged or enlarged but then connate and adnate to perianth only at base .... 53.

39.   Leaves broadly linear, 5 (8)–20 mm broad, often crescent-shaped. Perianth ovate-campanulate; as a result, tips of pink or greenish-yellow perianth lobes closed or campanulate, with lustrous pale yellow open perianth lobes; stamens longer than perianth ..... 40.

+   Leaves linear, up to 5 mm broad (if broader, perianth campanulate with pinkish perianth lobes; if with yellow lobes, stamens shorter than perianth), cylindrical, semicylindrical or convoluted bristle-like .... 43.

40.   Pedicels with numerous bracts at base; perianth lobes greenish-yellow. Leaves gradually narrowing towards tip, subacute. Bulb single, sessile on short rhizome .... 50. **A. obliquum** L.

+   Pedicel without bracts at base. Leaves hardly narrowing, broadly rounded at tip. 1–2 (4) bulbs sessile on rhizome .... 41.

41.   Perianth ovate-campanulate; lobes pink, opaque or slightly lustrous, not transparent; filaments enlarged at base .... 42.

+   Perianth campanulate; lobes pale yellow, lustrous, transparent; filaments not enlarged .... 46. **A. platystylum** Rgl.

42.   Filaments of inner stamens twice broader than outer at base. Capsule 2/3 length of perianth .... 33. **A. carolinianum** DC.

+   Filaments of inner stamens barely broader than outer at base. Capsule as long as perianth .... 32. **A. blandum** Wall.

43 (39). Stamens shorter than perianth; inner filaments enlarged for not less than 1/2 length .... 44.

+   Stamens longer than perianth; inner filaments not enlarged or barely enlarged at very base; rarely, enlarged for 1/5–1/4 length .... 48.

44.   Leaves filiform. Inflorescence fasciculate, few-flowered; pedicels with bracts at base .... 45.

+   Leaves flat. Inflorescence globose, many-flowered, pedicels without bracts at base .... 46.

45.   Bulbs oblong- or elongate-ovoid, with cream or yellowish-brown tunic. Pedicels usually twice or more longer than perianth; perianth lobes linear-lanceolate, 6–7 mm long, pink, with darker coloured nerve .... 47. **A. setifolium** Schrenk.

+ Bulbs ovate or oblong-ovate, with chestnut-coloured tunic. Pedicels usually shorter than perianth or as long, rarely longer; perianth lobes lanceolate or oblong-lanceolate, 5.5–6.5 (8) mm long, albescent, with dark violet nerve .... 41. **A. korolkovii** Rgl.

46. Filaments of stamens not enlarged; perianth lobes yellow, lustrous, transparent. Leaves 5–7 mm broad, crescent-shaped .... 35. **A. chrysocephalum** Rgl.

+ Filaments of inner stamens considerably enlarged. Leaves up to 3 mm broad .... 47.

47. Perianth lobes pale pink, linear-lanceolate. Leaves glabrous, clustered at scape base, with very short underground sheaths.... .... 44. **A. pevtzovii** Prokh.

+ Perianth lobes yellow, narrowly oblong-ovoid. Leaf sheaths and blades (along nerves on under surface) scabrous; leaf sheath elongate, underground .... 42. **A. megalobulbon** Rgl.

48 (43). Spathe with 1–2 cm long beak .... 49.

+ Spathe with very short, almost indistinct beak .... 51.

49. Perianth lobes rounded at tip (outer ones sometimes emarginate), yellowish but purple-pink in upper part and middle, sometimes wholly pinkish. Leaf blades semicylindrical or cylindrical, glabrous; sometimes diffusely and highly serrulate at base only; leaf sheaths glabrous. Bulb tunic golden-brown, lustrous ....49. **A. tianschanicum** Rupr.

+ Perianth lobes narrowing abruptly at tip into short recurved cusp. Leaf sheaths and blades glabrous or scabrous. Bulb tunic brown, opaque .... 50.

50. Leaf sheaths scabrous; leaf blades flat, serrulate along margin, veins projecting. Perianth lobes pale yellow or albescent, sometimes slightly pinkish .... 43. **A. petraeum** Kar. et Kir.

+ Leaf sheath glabrous; leaf blades semicylindrical, glabrous or sometimes diffusely and strongly serrulate at base. Perianth lobes purple in upper half, greenish below .... 38. **A. globosum** M.B. ex Redouté.

51. Inflorescence fasciculate, few-flowered; pedicels 2–3 times longer than perianth. Leaves filiform, about 0.5 mm broad. Perianth lobes pinkish-purple; filaments of stamens not enlarged, somewhat longer than perianth .... 48. **A. subtilissimum** Ledeb.

+ Inflorescence globose or subglobose, many-flowered. Leaves flat, 1–6 mm broad. Perianth lobes pink; filaments of stamens enlarged for 1/5–1/4 length, 1.5–2 times longer than perianth .... 52.

52. Bulbs 0.5–1 cm in diam. Midportion of scape 2–3 mm in diam. Leaves 1–3 mm broad, flat or sulcate. Pedicels thickened, shorter than perianth or equal; perianth lobes pale pink, nearly white (in herbarium) .... 40. **A. kaschianum** Rgl.

+ Bulbs 1.5–2 cm in diam. Midportion of scape 5–7 mm in diam. Leaves 5–6 mm broad, flat. Pedicels slender, 1.5–2 (3) times longer than perianth; perianth lobes pink .... 39. **A. hymenorrhizum** Ledeb.

53 (38). Stamens shorter than perianth or, rarely, equal .... 54.

+ Stamens longer than perianth, exserted .... 63.

54. Leaves 5–7 mm broad, falcate. Perianth lobes yellow, lustrous, transparent; filaments of stamens not enlarged............................ .... 35. **A. chrysocephalum** Rgl.

+ Leaves not more than 3 mm broad, erect. Perianth lobes of different colour; if yellow, neither lustrous nor transparent; filaments of inner stamens enlarged .... 55.

55. Flowers sky-blue; enlarged part of filaments of inner stamens with auriculate processes on both sides. Leaves 3 mm broad. (Plant from South. Tibet.) .... 31. **A. tibeticum** Rendle.

+ Flowers not sky-blue; filaments of stamens without auriculate processes. Leaves 1–2 (3) mm broad .... 56.

56. Perianth lobes (6) 7–8.5 mm long; filaments of inner stamens without teeth .... 57.

+ Perianth lobes (3) 3.5–4.5 mm long; if 5–6 mm long, filaments of inner stamens usually toothed .... 60.

57. Perianth lobes 6–7 mm long, elongate-ovate, with acute tip .... 21. **A. angulosum** L.

+ Perianth lobes (6) 7–8.5 mm long, elliptical, with broadly rounded tip .... 58.

58. Flowers dark purple, 3–5; pedicels shorter than perianth; style very shortly tripartite .... 62. **A. forrestii** Diels.

+ Flowers pinkish-lilac or white, sometimes pinkish, many; pedicel as long as perianth or 1.5–2 times longer; style undivided ...... 59.

59. Flowers pinkish-lilac, early shedding; filaments of inner stamens enlarged for 1/2 length, outer ones not enlarged............................ .... 25. **A. mongolicum** Rgl.

+ Flowers white, sometimes pinkish; filaments of inner stamens enlarged almost throughout length; outer filaments also enlarged, but twice narrower than inner .... 65. **A. caespitosum** Siev. ex Bong. et Mey.

60 (56). Pedicels unequal, often 0.5–3 cm long in same inflorescence, very rarely nearly equal; perianth lobes elliptical or oblong-elliptical, broadly rounded or suberect at tip, albescent or pinkish, with cuneate thickening of perceptibly denser texture at centre from base upward than elsewhere on surface; filaments of inner stamens without teeth .... 22. **A. anisopodum** Ledeb.

+ Pedicels equal or sometimes nearly equal; perianth lobes without cuneate thickening at centre; filaments of inner stamens with teeth, very rarely without teeth .... 61.

61. Perianth lobes pink or dark pink, outer ones oblong-elliptical or oblong-ovate; filaments of inner stamens usually with teeth. Inflorescence subglobose, rarely fasciculate, fairly compact. Bulb tunic blackish or greyish, fibrous .... 62.

+ Perianth lobes albescent or pinkish, outer ones round-elliptical; filaments of inner stamens without teeth. Inflorescence fasciculate or subglobose, lax, few-flowered. Bulb tunic albescent or pale brown, almost undivided. Very rare plant ....
.... 30. **A. tenuissimum** L.

62. Pedicels shorter than perianth, equal or slightly longer; filaments of inner stamens enlarged for 2/3 or 3/4 length. Inflorescence with not more than 12 flowers ....
.... 23. **A. bidentatum** Fisch. et Prokh.

+ Pedicels 2–3 times longer than perianth; filaments of inner stamens enlarged for 1/2 or 2/3 length. Not less than 20 flowers in inflorescence .... 24. **A. dentigerum** Prokh.

63 (53). Bulb tunic very strongly fibrous, light brown, sometimes rust-brown. Filaments of inner stamens enlarged. Roots several, thickened, subfuniform .... 64.

+ Bulb tunic almost undivided, very rarely fibrous but then filaments of stamens not enlarged. Roots not many, not thickened
.... 65.

64. Perianth lobes pink or pale pink (sometimes yellowish in herbarium), oblong-ovate, subobtuse or subacute, 4–5 (6) mm long; filaments of inner stamens enlarged for 1/2 of length, with teeth .... 13. **A. polyrhizum** Turcz. ex Rgl.

+ Perianth lobes pinkish-violet, oblong-lanceolate, acute or shortly acuminate, 6–7.5 mm long; filaments of inner stamens enlarged for 2/3 or 3/4 length, with teeth or sometimes without teeth ......
.... 19. **A. subangulatum** Rgl.

65. Leaves (5) 7–15 mm broad. Filaments of stamens not enlarged or slightly enlarged only at very base .... 66.

+ Leaves 1–3 mm broad; if broader, 5–8 (10) mm, filaments of inner stamens greatly enlarged for 1/2 or 2/3 length .... 67.

66. Perianth lobes pale yellow, transparent, oblong-elliptical, 7–8 mm long. Bulbs oblong-ovoid, up to 4 cm in diam ....
.... 46. **A. platystylum** Rgl.

+ Perianth lobes pink, not transparent, lanceolate or linear-lanceolate, (5) 6–8 mm long. Bulbs elongated-cylindrical ....
.... 45. **A. platyspathum** Schrenk.

67. Filaments of inner stamens enlarged for not less than 1/3 length. ....68.
+ Filaments of inner stamens not enlarged or barely enlarged at very base ....70.
68. Inflorescence fasciculate, lax, few-flowered; pedicels 2–2.5 times longer than perianth; perianth lobes 5.5–6 mm long, pinkish-violet, ovate, narrowed toward tip, subacute; filaments of stamens enlarged for 1/2 of length. Leaves longitudinally convoluted, 1–2 mm broad ....27. **A. rubens** Schrad. ex Willd.
+ Inflorescence globose or, rarely, subglobose, usually compact, many-flowered ....69.
69. Leaves (2) 4–8 (10) mm broad, flat. Inflorescence globose, (2) 3–5 cm in diam.; pedicels 2–3 (4) times longer than perianth; perianth lobes generally oblong or oblong-lanceolate; enlarged part of filaments of inner stamens usually almost as long as perianth lobes and occasionally with short teeth. Bulbs prominently developed, oblong-ovoid, up to 2 cm thick. Plant 30–60 cm tall .. .. 28. **A. senescens** L.
+ Leaves 0.5–2 mm broad, flat or longitudinally convoluted. Inflorescence globose or subglobose, up to 2 cm in diam.; pedicels as long as perianth or slightly longer; perianth lobes generally ovate or broadly ovate; enlarged part of filaments of inner stamens usually 1/2 length of perianth lobes, without teeth. Bulbs smaller. Plant 10–25 cm tall .... 26. **A. prostratum** Trev.
70 (67). Inflorescence fasciculate; perianth pink. Bulb tunic fibrous .... 37. **A. consanguineum** Kunth.
+ Inflorescence globose, very compact; perianth yellow, sometimes pinkish. Bulb tunic almost undivided ....71.
71. Perianth lobes golden-yellow, lustrous, transparent, elliptical or oblong-elliptical, 5.5–6.5 mm long; spathe greyish, with numerous black dots. Leaves 2–3 mm broad .... 34. **A. chrysanthum** Rgl.
+ Perianth lobes pale yellow, opaque, not transparent, broadly elliptical, 4–5 mm long; spathe albescent, without dots. Leaves up to 1.5 mm broad..................................... 29. **A. stellerianum** Willd.
72 (6). Bulbs not developed; numerous thickened fleshy roots seen. Scape base surrounded by sheaths of preceding year's leaves laciniated into parallel fibres. Leaves 3–4 mm broad. Perianth white; filaments of stamens not enlarged. (Plant from South. Tibet.) ....66. **A. fasciculatum** Rendle.
+ Bulbs well-developed, solitary ....73.
73. Lower part of scape foliate; leaves clustered at scape base, with underground sheaths and leaf blades emerging nearly at same level...74.

+ Scape foliate for most part, with surface sheaths and leaf blades emerging at different levels ....76.
74. Perianth broadly campanulate, pinkish-purple; perianth lobes elliptical or broadly elliptical, 8–10 mm long, 5–6 mm broad; stamens 1/2 length of perianth; style shortly tripartite ....
.... 79. **A. oreophilum** C.A. Mey.
+ Perianth stellate; perianth lobes narrowly linear, 5–6 mm long, 1.5–2 mm broad; stamens almost equal to perianth; style undivided ....75.
75. Flowers dark purple; notches between free (not connate) parts of filaments of stamens cuneate and usually narrow ....
.... 81. **A. robustum** Kar. et Kir.
+ Flowers pale pinkish-violet; notches between free parts of filaments of stamens rounded or suberect, broad ....
.................................80. **A. decipiens** Fisch. ex Schult. et Schult. f.
76. Spathe with long (up to 4.5 cm) beak; flowers yellow ....
.... 78. **A. stamineum** Boiss.
+ Spathe with short (not longer than 0.5 cm) beak ....77.
77. Stamens shorter than perianth, generally by 1.5–2 times; inner filaments enlarged for 2/3 or 3/4 length and twice broader than outer ones, with 2 teeth above or without teeth ....78.
+ Stamens generally longer than perianth, sometimes almost twice, rarely equal or 1/4 shorter; filaments enlarged only at very base or for not more than 1/2 of length or not enlarged at all; inner filaments almost indistinguishable from outer in breadth, without teeth above ....81.
78. Pedicels short, equal to perianth or 1.5–2 times longer, without or with few bracts at base; stamens 1.5–2 times shorter than perianth; inner filaments without or with very short teeth above
....79.
+ Pedicels elongated, 3–5 times longer than perianth, with numerous bracts at base resembling tiny tomentum; stamens slightly shorter than perianth, almost as long; inner filaments with distinct narrow teeth above. Flowers, sky–blue to blue ....
.... 67. **A. caesium** Schrenk.
79. Flowers 5–7 mm long; perianth lobes lustrous, pink, with mid-nerve somewhat darker than background; filaments of inner stamens without or with short teeth ....80.
+ Flowers 4.3–5 mm long; perianth lobes pink, with distinct nerve considerably darker than background; filaments of inner stamens without teeth .... 71. **A. jacquemontii** Kunth.
80. Filaments of inner stamens without teeth, cuneately narrowing upward.. 75. **A. sairamense** Rgl.

+ Filaments of inner stamens with short teeth ....
.... 76. **A. schoenoprasoides** Rgl.

81. Pedicels short, as long as perianth; spathe with solitary distinct midnerve (lateral nerves very faint) and with numerous fine black dots; filaments of stamens not enlarged. Flowers pink ....
.... 70. **A. glomeratum** Prokh.

+ Pedicels usually 2–5 times longer than perianth; distinctly manifest lateral nerves seen on spathe apart from midnerve; filaments of stamens enlarged for 1/4 or 1/2 length, rarely not enlarged .... 82.

82. Leaves 2–4 mm broad, borne in lower part of scape; leaf blades emerging from scape nearly at same level; distance from scape base to point of emergence of upper leaf not more than 7 cm. Flowers violet-pink .... 77. **A. tanguticum** Rgl.

+ Leaves usually 0.5–2 mm broad; extremely rarely up to 4 mm broad. Scape foliate for major part; leaf blades emerging from scape at different levels; distance from scape base to point of emergence of upper leaf considerably greater .... 83.

83. Scape thickened, inflated, hollow. Leaves narrow, fistular. Bulb tunic somewhat crimped, with sharp impressed nerves. Pedicels with numerous white bracts at base forming tiny tomentum; perianth lobes greenish or albescent, outer ones rugulose and scabrous and inner slightly rugose and emarginate above; filaments of stamens not enlarged .... 74. **A. sabulosum** Stev. ex Bge.

+ Scape slender, solid. Leaves flat or sulcate. Nerves (if present on bulb tunic) not impressed. All perianth lobes glabrous and inner ones not emarginate; filaments of stamens enlarged; pedicels with numerous or few bracts at base .... 84.

84. Pedicel base with numerous white bracts (as though with tiny tomentum). Umbel sometimes with bulbils .... 85.

+ Pedicel base with few bracts. Umbel without bulbils .... 86.

85. Filaments of stamens enlarged roughly for 1/5 length and enlarged part nearly completely adnate to perianth lobes; stamens slightly shorter than perianth or as long; flowers albescent or pinkish. Umbel largely fasciculate or semiglobose, without bulbils. Leaves filiform, sulcate, 0.5–1 mm broad ....
.... 69. **A. delicatulum** Siev. ex Schult. et Schult. f.

+ Filaments of stamens enlarged for 1/2 of length and 1/2 of enlarged part adnate to perianth lobes; stamens 1/4 longer than perianth; flowers dark pink. Umbel subglobose or globose often with bulbils. Leaves linear, 2–4 mm broad ....
.... 72. **A. macrostemon** Bge.

86. Perianth lobes blue, their entire surface of same consistence; stamens as long as perianth or slightly longer; spathe pergamentaceous, rigidulous, cream-coloured, almost non-transparent, with achromatic nerves. Leaves triquetrous-sulcate, (1) 2–4 mm broad .... 68. **A. coeruleum** Pall.

+ Perianth lobes pink, with cuneate thickening at centre above base with denser consistence than rest of surface; stamens slightly or almost twice longer than perianth; spathe scarious, soft, achromatic, nearly transparent, usually with purple nerves. Leaves filiform or narrowly linear, 0.5–1.5 (2.5) mm broad ....
.... 73. **A. pallassii** Murr.

Subgenus RHIZIRIDEUM (Koch) Wendelbo

Section **Anguinum** G. Don ex Koch

1. **A. ovalifolium** Hand.-Mazz. in Anzeig. Akad. Wiss. (Wien) 60 (1924) 101. —*A. victorialis* auct. non L.: Forbes and Hemsley, Index Fl. Sin. 3 (1905) 126, p.p., quoad pl. tangut. et seczuan.; Walker in Contribs. U.S. Nat. Herb. 28 (1941) 602.

Described from South-west China (Yunnan). Type in Vienna (W).

Humid areas near springs and small streams in deciduous forests in lower part of mountain belt.

IIIA. **Qinghai:** *Nanshan* (South Tetung mountain range, in lower part of forest belt, Aug. 1872—Przew.; same locality, humid areas near springs in deciduous forests, 2200 m, Aug. 7, 1880—Przew.; "Shui Mo Kou, near Lien Cheng, in woods, 1923, Ching"—Walker, l.c.).

General distribution: China (South-west).

2. **A. prattii** C.H. Wright in J. Linn. Soc. London (Bot.) 36 (1903) 124; id. in Forbes and Hemsley, Index Fl. Sin. 3 (1905) 124.

Described from South-west China. Type in London (K).

Upper mountain belt.

IIIB. **Tibet:** *Weitzan* (Yangtze basin, Donra area along Khichu river, in thickets, 3900 m, July 16, 1900—Lad.).

General distribution: China (South-west).

A. victorialis L. Sp. pl. (1753) 295; Rgl. in Acta Horti Petrop. 3, 2 (1875) 170; Franch. Pl. David. 1 (1884) 306; Forbes and Hemsley, Index Fl. Sin. 3 (1905) 126, p.p.; Kryl. Fl. Zap. Sib. 3 (1929) 629; Vved. in Fl. SSSR, 4 (1935) 141; Kitag. Lin. Fl. Mansh. (1939) 132; Grub. Konsp. fl. MNR (1955) 94.

Described from Europe. Type in London (Linn.).

Coniferous forests, forest fringes and meadows, moist slopes of ravines.

Occurrence of species possible in regions adjoining Nor. Mongolia (Cen. Khalkha, East. Mong.).

General distribution: Europe, Caucasus, West. Sib.; East. Sib. (including Sayans), Far East, Nor. Mong. (Hent., Mong.-Daur.), China (Dunbei, Nor.), Korean peninsula, Japan, North America.

Section **Reticulato-bulbosa** R. Kam.

3. **A. amphibolum** Ledeb. Fl. alt. 2 (1830) 5; ej. Fl. Ross. 4, 1 (1852) 179; Rgl. in Acta Horti Petrop. 3, 2 (1875) 166; Sap. Mong. Alt. (1911) 388; Kryl. Fl. Zap. Sib. 3 (1929) 625; Vved. in Fl. SSSR, 4 (1935) 152; Grub. Konsp. fl. MNR (1955) 92, p.p., excl; pl. Cischubsugul.; Fl. Kazakhst. 2 (1958) 146. — Ic.: Ledeb. Ic. pl. fl. ross. 6 (1833) tab. 357.

Described from Altay. Type in Leningrad.

Rocky slopes in high mountains, rarely in montane steppes and upper part of forest belt.

IA. Mongolia: *Khobd.* (south. tributaries of Kharkhira, in alpine belt, July 23, 1879—Pot.); *Mong. Alt.* (in Datu-Daban pass, July 17, 1894—Klem.; Tamchi somon, nor. trail of Bus-Khairkhan mountain range, gully in upper third of trail, July 17; Adzhi-Bogdo mountain range, sheep's fescue montane steppe, Aug. 7—1947, Yun.; Bulugun river basin, left tributary of Ulyastei-Gol, on rocks near upper forest boundary, June 28, 1973—Golubkova and Tsogt; "Dzhyumaly, Ukok, Ulan-Daba, Tsagan-Aksu, Kulagash; in alpine belt, 1905–1909"—Sap. l.c.); *Gobi-Alt.* (Ikhe-Bogdo mountain range, Bityuten-Ama creek valley, slopes of creek valley and alpine belt, Aug. 12, 1927—Simukova; nor. part of Dzun-Saikhan mountain range, upper third of nor. slope, June 19, 1945—Yun.).

IIIA. Qinghai: *Nanshan* (in Nanshan alps, 1879—Przew., without reference to exact site of occurrence or date of collection).

General distribution: West. Sib. (Alt.), East. Sib. (south.), Nor. Mong. (Fore Hubs., Hent., Hang.), China (Altay).

Note. Closely related to *A. strictum* Schrad., *A. lineare* L. and *A. schrenkii* Rgl., differing from them in complex of variable characters. Differs from *A. strictum* (closest relative) in very dense capitate inflorescence and short (shorter than or equal to perianth) pedicels which, however, may be as long as perianth in *A. strictum* (as well as in *A. lineare*). Differs from *A. schrenkii* in presence of teeth on enlarged filaments of inner stamens (teeth sometimes absent). From all these species, *A. amphibolum* differs in (usually) long styles, strongly exserted from perianth.

4. **A. clathratum** Ledeb. Fl. Alt. 2 (1830) 18; Rgl. in Acta Horti Petrop. 3, 2 (1875) 173; Kryl. Fl. Zap. Sib. 3 (1929) 627; Vved. in Fl. SSSR, 4 (1935) 145; Grub. Konsp. fl. MNR (1955) 92. —*A. ubsicolum* Rgl. l.c. 10, 1 (1887) 342.

Described from Altay. Type in Leningrad.

Steppes, rubble and rocky slopes, rocks; lower and middle mountain belts.

IA. Mongolia: *Khobd.* (Altyntschetsche, June 9, 1870—Kalning); *Mong. Alt.* (Dain-Gol lake, arid slopes, July 12, 1908—Sap.; Khara-Dzarga hills, Sakhirsala river valley, south. rubble slopes, Aug. 23 and Aug. 24, 1930—Pob.); *Depr. Lakes* (environs Ubsu-Nur lake south of Ulangom, Sept. 4, 1879—Pot.); *East. Gobi* (Shabarakh Usu, dry wash at 1140 m, 1925—Chaney).

IIA. Junggar: *Cis-Altay* (Kandagatai, in shrubs, Sept. 14, 1876—Pot.; Qinhe [Chingil'], 800 m, No. 1259, Aug. 5, 1956—Ching; 20 km from Shara-Sume settlement on Kran river, scrub meadow steppe, No. 1141, July 7, 1959—Yun. and I.-F. Yuan').

General distribution: West. Sib. (south-east including Altay), East. Sib. (south.), Nor. Mong. (Hang.).

Note. Closely related to *A. lineare* and *A. strictum* from which it differs in nearly cylindrical or cylindrical-sulcate, very narrow, glabrous (but not flat and costate) leaf blades. Teeth on enlarged filaments of inner stamens as in *A. lineare*: acute, erect (sometimes teeth absent). In appearance, *A. clathratum* closely resembles *A. strictum* but differs in more compact inflorescence.

5. **A. cyaneum** Rgl. in Acta Horti Petrop. 3, 2 (1875) 174 and 10, 1 (1887) 346, excl. var. *brachystemon*; Kanitz in Széchenyi, Keletazsiai utjanák, 2 (1891) 59; id. in Széchenyi, Wissensch. Ergebn. 2 (1898) 734; Forbes and Hemsley, Index Fl. Sin. 3 (1905) 122; Hao in Bot. Jahrb. 68 (1938) 587; Walker in Contribs, U.S. Nat. Herb. 28 (1941) 602. —Ic.: Rgl. l.c. 10, 1, tab. IV, fig. 3, 3, c.

Described from Qinghai (Nanshan). Type in Leningrad. Plate II, fig. 6.

River-banks, alpine meadows; upper part of forest and alpine belts.

IA. Mongolia: *Alash. Gobi* ("Ho Lan Shan, on shaded, mossy forest floor, Ching, 1923"—Walker, l.c.).

IC. Qaidam: *hilly* (Dulankhit temple, in spruce and juniper forests, 3300 m, Aug. 8, 1901—Lad.).

IIIA. Qinghai: *Nanshan* (Rako-Gol river valley between Nanshan and Donkyr mountain ranges, alpine belt, 3000–3300 m, July 22; South Tetung mountain range, forest belt, 2550 m, Aug. 3—1880, Przew.; North Tetungsk mountain range, river-banks, 2400–2700 m, Aug. 13, 1880—Przew., typus!; bank of Tetung river above Yukan'dzen, Aug. 8, 1890—Gr.-Grzh.; South Kukunor mountain range, Bain-Gol area, in meadow, 3300 m, July 26, 1894—Rob.; Yangtze river-basin, Nruchu area, along slopes, rocks, sandy river banks, 3500 m, July 25, 1900—Lad.); *Amdo* ("In jugi Kvetechiensis latere bor. init. Aug. 1879, Széchenyi"—Kanitz, l.c.; "In der Nähe des Klosters Taschinsze zwischen Steppenpflanzen, Aug. 26, 1930"—Hao, l.c.).

IIIB. Tibet: *Weitzan* (on way from Dzhagyn-Gol river to Razboini lake, 4050 m, Aug. 26; Alaknor-Gol river, Aug. 11—1884, Przew.; Mekong river-basin, Tszachu river, Nov. 1900—Lad.; "Amne-Matchin, bis 4500 m, Sept. 2, 1930"—Hao, l.c.).

General distribution: China (North-west, Cent., South-west: Sichuan'), Himalayas (east.).

6. **A. eduardii** Stearn in Herbertia, 11 (1944) 102, in adno.; Vved. in Opred. rast. Sr. Azii [Key to Plants of Mid. Asia], 2 (1971) 56. —*A. fischeri* Rgl. in Acta Horti Petrop. 3, 2 (1875) 161 and 10, 1 (1887) 342, non Roem. et Schult. (1830); Kryl. Fl. Zap. Sib. 3 (1929) 624; Vved. in Fl. SSSR, 4 (1935) 145; Grub. Konsp. fl. MNR (1955) 93; Fl. Kazakhst. 2 (1958) 144.

Described from Altay. Type in Leningrad.

Rubble and rocky slopes and rocks.

IA. **Mongolia:** *Khobd.* (east. foothills of Kharkhir mountain range, hillocky area, Aug. 22, 1944—Yun.; Achit-Nur lake, hillocky area east of Bukhu-Muren somon, nor. of lake, on rocks, July 15, 1971—Grub. et al.); *Mong. Alt.* (Shiverin-Gol valley south of Kobdo, arid slopes, July 20, 1906—Sap.; 40 km nor. of Kobdo along road to Tsagan-Nur, smoothened hillocky area, Aug. 7, 1945—Yun.); *Cent. Khalkha* (Bichikte-Dulan-Khada mountains, on rocky sites, Aug. 31, 1925; Bichikte mountains, ridge east of Khangai mountain range near Toly river meander, Aug. 28, 1926—Gus.; Ikhe-Tukhum-Nor lake environs, Ongon-Khairkhan mountains, June 1926—Zam.); *Depr. Lakes* (Kharkhira river valley, south. slopes, Aug. 3, 1931—Bar.; south-west. Fringe of Uryuk-Nur lake-basin, feather-grass desert steppe along knoll slopes, July 30, 1945—Yun.; Ubsa-Nur, hillocky area south-east of Ulangom, stepped desert along knoll slope, July 19, 1971—Grub. et al.; right bank of Khungui, south-east of Ikhe-Margats ula offshoots, along rocks, Aug. 18, 1972—Grub. et al.); *Gobi-Alt.* (Khurkhu mountain range, Ikhe-Nomogon hills, July 22, 1930—Simukova; Bain-Tsagan mountains, near rocks, Aug. 5; Tszolin mountain, rocky slope, Aug. 8; Dundu-Saikhan mountains, rocky slopes, Aug. 17; Dzun-Saikhan mountains, rocky slopes, Aug. 24; Barun-Saikhan mountains, near rocks, Sept. 20—1931, Ik.-Gal.; Bain-Tsagan mountain range, mountain slopes from trail to upper belt, July-Aug. 1933—Khurlat and Simukova; Noyan somon, Noyan-Bogdo-Ula 4–5 km south of somon, rocky slopes of mountains, feather-grass—desert steppe, July 25; west. extremity of Baga-Bogdo mountain range, rubble-sand along lower part of trails, Sept. 17; Baga-Bogdo mountain range, middle belt, forb-wormwood steppe, Sept. 18—1943, Yun.; Severei somon, Nemegetu-Nuru mountain range, peak of Khara-Obo, on peak and along rocky slope, Aug. 7, 1948—Grub.); *West. Gobi* ("Atas-Bogdo"—Grub. l.c.); *Alash. Gobi* (Alashan mountain range, Yamata gorge, west. slope, May 1, 1908—Czet.).

IIA. **Junggar:** *Jung. Gobi* (nor.: sand along south-east. bank of Ulyungur lake, Aug. 15, 1876—Pot.).

General distribution: Jung.-Tarb.; Nor. Mong. (Hang.).

7. **A. flavidum** Ledeb. Fl. alt. 2 (1830) 7; ej. Fl. Ross. 4, 1 (1852) 179; Rgl. in Acta Horti Petrop. 3, 2 (1875) 168 and 10, 1 (1887) 344; Sap. Mong. Alt. (1911) 388; Kryl. Fl. Zap. Sib. 3 (1929) 628; Vved. in Fl. SSSR, 4 (1935) 146; Fl. Kazakhst. 2 (1958) 144; Vved. in Opred. rast. Sr. Azii [Key to Plants of Mid. Asia], 2 (1971) 56. —*A. leucocephalum* auct. non Turcz. ex Ledeb.: Grub. Konsp. fl. MNR (1955) 93, p.p., quoad pl. chang. —**Ic.:** Ledeb. Ic. pl. fl. ross. 4 (1833) tab. 362.

Described from Altay. Type in Leningrad.

Meadows, alpine and subalpine belts.

IA. **Mongolia:** *Mong. Alt.* (bank of Khartsiktei river, above forest boundary, July 27, 1898—Klem.; upper Kharagaitu-Gol, left tributary of Bulugun, larch forest, Aug. 24; upper Indertiin-Gol, subalpine steppe, Aug. 25—1947, Yun.); *Gobi-Alt.* (Gurban-Saikhan, middle belt of Dundu-Saikhan mountains, steppe below pass, rocky slope, July 22, 1943—Yun.).

IIA. **Junggar:** *Cis-Alt.* (Kairta river valley, alpine tundra, July 16, 1908—Sap.; "Ul'kun-Kairty and Kanas rivers, 1905-1909"—Sap.; l.c.; Qinhe [Chingil'], No. 1082, Aug. 9, 1956—Ching); *Tarb.* (Dachen [Chugchak], 1830 m, No. 41601, Aug. 13, 1957—Kuan); *Jung. Alatau* (Toli district, on slopes, No. 2531, Aug. 6, 1957—Kuan); *Tien Shan* (Balun'tai to north of Bagrash-Kul' lake, subalpine meadow, No. 1062,

Aug. 6, 1957—Kuan); *Jung. Gobi* (east.: Bulugun somon, Baitak-Bogdo-Nuru mountain range, left creek valley of Ulyastu-Gol gorge, on talus along southern slope, 2000 m, Sept. 17, 1948—Grub.).

General distribution: Jung.-Tarb.: West. Sib. (Alt.), East. Sib. (south.), Nor. Mong. (Hang.).

Note. *A. flavidum* is closely related to *A. leucocephalum* and very similar to it in appearance, especially in dry state (in herbarium). The two plants have similar capitate inflorescence and yellow or pale yellow to albescent perianth lobes which are sometimes pinkish-violet at centre and tip. The difference between the two is that leaves of *A. leucocephalum* are cylindrical or semicylindrical and glabrous, while those of *A. flavidum* are flat with slender, highly serrulate margin (leaf breadth in both species varies from 1 to 5 mm). Moreover, filaments of inner stamens of *A. leucocephalum* versus those of *A. flavidum* bear distinctly longer teeth, sometimes reaching the tip of the perianth lobes.

8. **A. kansuense** Rgl. in Acta Horti Petrop. 10, 2 (1889) 690; Baker in Curtis's Bot. Mag. 49, 580 (1893) tab. 7290; Diels in Futterer, Durch Asien, 3 (1903) 7; Forbes and Hemsley, Index Fl. Sin. 3 (1905) 123; Hao in Bot. Jahrb. 68 (1938) 587; Walker in Contribs. U.S. Nat. Herb. 28 (1941) 602. — *A. cyaneum* var. *brachystemon* Rgl. in Acta Horti Petrop. 10, 1 (1887) 346. — Ic.: Rgl. l.c. (1887) tab. IV, fig. 3, a (sub nom. *A. cyaneum* Rgl.); Curtis's Bot. Mag. tab. 7290.

Described from Qinghai (Nanshan). Lectotype in Leningrad. Plate II, fig. 3. Alpine meadows.

IIIA. **Qinghai:** *Nanshan* (along Tetung river, July 28, 1872; south. slope of South Tetung mountain range, alpine meadow, July 30, 1880—Przew., lectotypus! Nor. Tetung mountain range, moist alpine meadow, 3000 m, Aug. 15—1880, Przew.; on way from Alashan to Kukunor lake, Sangyn area, on south-east. slope in middle belt, Aug. 12, 1908—Czet.; "Nan-Schan an der Wasserscheide bei Schalakuto [undated] Futterer"—Diels, l.c.; "Kükenur-Gebiet: Nordfluss der Korallenkalkberge westlich des Hoang-ho bei Lager XVII [undated] Futterer"—Diels, l.c.; "La Ching Kou, 1923, Ching"—Walker, l.c.; "Schang-wu-chuang, unweit Sinning-fu, Aug. 3, 1930"—Hao, l.c.).

IIIB. **Tibet:** *Weitzan* (Tibet borealis, 1884—Przew. [without precise date or locality]; "Amne-Matchin, 4500 m, Sept. 2, 1930"—Hao, l.c.).

General distribution: China (North-west, South-west: Sichuan).

9. **A. leucocephalum** Turcz. ex Ledeb. Fl. Ross. 4, 1 (1852) 179; Vved. in Fl. SSSR, 4 (1935) 146; Kitag. Lin. Fl. Mansh. (1939) 131; Gordeev and Jernakov in Acta pedol. sinica, 2, 4 (1954) 277; Grub. Konsp. fl. MNR (1955) 93. —*A. flavo-virens* Rgl. in Acta Horti Petrop. 10, 1 (1887) 344, 294; Forbes and Hemsley, Index Fl. Sin. 3 (1905) 122. —*A. schischkinii* Sobol. in Sist. zam. Gerb. Tomsk. univ. 1–2 (1949) 10; Grub. Konsp. fl. MNR (1955) 94. —Ic.: Rgl. l.c. tab. 8, fig. 1 (sub nom. *A. flavo-virens*).

Described from East. Siberia. Type in Leningrad.

Steppe rubble and melkozem slopes of mountains, sand, knolls; plains, foothills and lower mountain belt.

36

IA. **Mongolia:** *Mong. Alt.* (Bulgan somon, upper Vayan-Gol on road to summer camp in somon, steppe slope, July 23, 1947—Yun.); *Cent. Khalkha* (pass through Tono-Ul mountains, July 21; Kerulen upper course, crossing on right bank of river opposite Tono-Ul mountains; Kerulen midcourse, Berlik mountain, on sand—1899, Pal.; Butszynkhe sand, Aug. 11, 1925—Gus.; Ikhe-Tukhum-Nor lake environs, Temeni-Ama gorge, June, 1926—Zam.; 50 km south-east of Ugapur camp along road from Ulan-Bator to Choiren—Sain-Shandu, feather-grass—wild rye steppe, July 23, 1941—Yun.; 59 km north of Munku-Khan somon along road to Undur-Khan, love-grass—wormwood solonchak lowland, July 29; Bain-Barat somon, 35-40 km south of Buralyn-Daba pass, wormwood—forb steppe on rubble ridge, Aug. 9—1962, Yun. and Dashnyam); *East. Mong.* (near Kharkhonte railway station, on sand, Aug. 20, 1902—Litw.; watershed of Ara-Dzhargalante and Uber-Dzhargalante rivers, on sand, 1925—Krasch. and Zam.; vicinity Baishintin-Sume, Ongon-Elis, Aug. 20, 1927—Zam.; Kerulen river below Tsetsen-Khan, arid slopes, Aug. 12, 1928—Tug.; Dariganga, Ongon-Elis sand, on sand knolls, Sept. 14, 1931—Pob.; 30 km north of Choibalsan, slopes of knolls, Aug. 12, 1954—Dashnyam); *Depr. Lakes* ("South. slopes of TannuOl"—Sobol. l.c.; Grub. l.c.), *Gobi-Alt.* (spur of Ikhe-Bogdo mountain range, July 10, 1926—E. Kozlova; Dundu-Saikhan mountain, along slopes, Aug. 17, 1931—Ik.-Gal.; west. extremity of Baga-Bogdo mountain range, upper part of trail, wormwood—forb desert steppe, Sept. 17, 1943—Yun.).

IIA. **Junggar:** *Cis-Alt.* (Sarbyuira river, grassy slopes, Sept. 25, 1876—Pot.).

General distribution: East. Sib. (south.), Nor. Mong., China (Dunbei).

Note. While describing *A. flavo-virens* Rgl. l.c., treated here as a synonym of *A. leucocephalum*, it was stated that this plant was collected by Przewalsky in Gansu province of North-west China. The Herbarium of the Komarov Botanical Institute, Academy of Sciences of the USSR (Leningrad) has only a single specimen which Regel designated as *A. flavo-virens*. This specimen, dated July 1887, was grown in the Botanical Garden. In all probability, it should be treated as a type of *A. flavo-virens*. The latter is identical to *A. leucocephalum* which, however, is not found in Gansu province. Evidently, the labels were mixed up in the garden and the plant grown and described as *A. flavo-virens* was neither collected in Gansu province nor perhaps by Przewalsky.

A. schischkinii Sobol. l.c. was described from Tuva Autonomous Soviet Socialist Republic (Eleges river valley) and cited by the author and V.I. Grubov (l.c.) for Mongolia (Great Lakes Depression: south. slope of Tannu-Ol mountain range). A study of an authentic specimen of this species (Tuva province, East. Tannu-Ol mountain range, Turgen' river valley, July 30, 1945, K.A. Sobolevskaja and A.A. Khor'kova) received from P.N. Krylov Herbarium (Tomsk University) revealed its similarity with *A. leucocephalum*.

The characters cited by Sobolevskaja which distinguish *A. schischkinii* from *A. leucocephalum* (leaves as long as stem or slightly shorter, subglobose and on globose inflorescence, canescent perianth lobes pinkish in upper part and filaments of stamens white) are variable and have been observed among specimens of *A. leucocephalum* throughout its distribution range. The colour of filaments of stamens—white or yellow—depends on the stage of plant flowering. In the authentic specimen of *A. schischkinii*, filaments are yellow in some inflorescences and white in others.

10. **A. lineare** L. Sp. pl. (1753) 295; Rgl. in Acta Horti Petrop. 3, 2 (1875) 166; Regel in Izv. Obshch. lyubit. estestvozn. antrop. etnogr. 21, 2 (1876) 82; Rgl. in Acta Horti Petrop. 10, 1 (1887) 344, p.p., excl. syn.; Forbes and Hemsley, Index Fl. Sin. 3 (1905) 123; Sap. Mong. Alt. (1911) 388; Kryl. Fl.

Zap. Sib. 3 (1929) 625, p.p., excl. var. *strictum* (Schrad.) Kryl.; Gr.-Grzh.
Zap. Mong. 3, 2 (1930) 807; Vved. in Fl. SSSR, 4 (1935) 150; Grub. Konsp.
fl. MNR (1955) 93, p. max. p., excl. pl. ex Alt. mong. et gobic.; Fl. Kazakhst.
2 (1958) 144; Vved. in Opred. rast. Sr. Asii [Key to Plants of Mid. Asia], 2
(1971) 56. —*C. amphibolum* auct. non Ledeb.: Grub. l.c., quoad pl. ex
Cischubsugul. —Ic.: Regel, l.c., 1876, tab. 13, figs. 1–5.

Described from Siberia. Type in London (Linn.).

Meadows and meadow steppes; up to alpine belt.

IA. **Mongolia:** *Mong. Alt.* (south. Altay, in lower course of Tatal river, July 8,
1877—Pot.; Bodonchin-Gol valley along right rocky slope of valley, 1600 m, June 30,
1973—Golubkova and Tsogt; "Tiekty mountain range, alpine meadow, July 18,
1903"—Gr.-Grzh.; "Tsagan-Gol; Upper Kobdo lake; Sumdairyk; Dain-Gol lake, 1905–
1909"—Sap., l.c.); *East. Mong.* (Huang He river valley, Muni-Ula hill, July 13, 1871—
Przew.; Ourato, July 1866, A. David; Khailar, July 6, 1901—Litw.); *East. Gobi*
(Mongolia, Gobi—Przew. [with no reference to specific locality or date of collection]).

IIA. **Junggar:** *Cis-Alt.* (15–20 km nor.-west of Shara-Sume, scrub steppe, July 7,
1959—Yun. and I.-F. Yuan'; 35 km south of Koktogai along road Ertai, on rocky slope
of scrub steppe, July 14, 1959—Yun.).

General distribution: Aralo-Casp., Fore Balkh., Jung.-Tarb.; Europe (European
USSR), West. Sib. (south.), East. Sib. (south.), Nor. Mong. (Fore Hubs., Hent., Hang.),
China (North).

11. **A. odorum** L. Mantissa, 1 (1767) 62; Rgl. in Acta Horti Petrop. 3, 2
(1875) 175 and 10, 1 (1887) 346; Franch. Pl. David. 1 (1884) 306; Forbes
and Hemsley, Index Fl. Sin. 3 (1905) 123; Kryl. Fl. Zap. Sib. 3 (1929) 630;
Gr.-Grzh. Zap. Mong. 3, 2 (1930) 807; Vved. in Fl. SSSR, 4 (1935) 163; Hao
in Bot. Jahrb. 68 (1938) 586; Kitag. Lin. Fl. Mansh. (1939) 131; Gordeev and
Jernakov in Acta pedol. sin. 2, 4 (1954) 277; Grub. Konsp. fl. MNR (1955)
93; Chen and Chou, Rast. pokrov r. Sulekhe [Vegetal Cover of Sulekhe
River] (1957) 92; Fl. Kazakhst. 2 (1958) 151; Fl. Tadzh. 2 (1963) 308. —*A.
tataricum* L. f. Suppl. (1781) 196; Danguy in Bull. Mus. nat. hist. natur. 20
(1914) 140; ? Paulsen in Hedin, S. Tibet, 6, 3 (1922) 89. —*A. potaninii* Rgl.
in Acta Horti Petrop. 6 (1879) 295.

Described from Siberia. Type in Stockholm (S)?, Plate III, fig. 6.

Steppes, arid solonetz meadows, steppe melkozem and rocky slopes of
mounds and mud cones, pebble beds; plains and foothills.

IA. **Mongolia:** *Mong. Alt.* ("Kobdo town environs, July 14, 1906"—Sap. l.c.); *Cen.
Khalkha* (upper course of Kerulen river, steppe near Bain-Erkhetu hills, July;
midcourse of Kerulen river, steppe near Shokhul' hill [undated]—1899, Pal.; Khukhu-
Khoshun, rocky slope, July 24, 1926—Lis.; Ikhe-Tukhum-Nor lake environs, May
1926—Zam.; Kerulen river 150 km below Tsetsen-Khan, arid slopes, Aug. 12, 1928—
Tug.; Sorgol-Khairkhan town 130 km south-west of Ulan Bator along old road to
Dalan-Dzadagad, in crevices among granite structures, July 15, 1943—Yun.; left bank
of Kerulen on Ulan Bator—Undur-Khan road, rocky slopes of knolls, July 22; 10 km
east of Idermeg khid along road, feather-grass—snakeweed steppe, Aug. 30—1949,
Yun.; 10 km west of Tuman-Delger somon, Marzyn-Khola area, in steppe, Aug. 8,

1956—Dashnyam); *East. Mong., Depr. Lakes* (nor.-east. extremity of Khirgiz-Nur basin, Barin-Ula knolls 27 km south-east of Undur-Khangai somon along road to Numuryg, trail on granite detritus, July 23, 1971—Grub. et al; "Ubsa-Nur, Khobdo"— Grub. l.c.); *East. Gobi* (375 miles south-east of Urga, Aug. 1927—A. Terekhovko); *Ordos* (Ordos, 1873—Przew.; near Tsaidaminchzhao monastery, Aug. 9, 1884—Pot.; between Dzaoliping and Shundantszen, Tszokul'-Qaidam area, Sept. 7, 1885—Pot.); *Khesi* ("Sulekhe river"—Chen and Chou, l.c.).

IB. **Kashgar:** *East.* (nor. extremity of Khami desert, Chanmo well, pebble steppe, 600 m, Aug. 18, 1895—Rob.).

IIA. **Junggar:** *Cis-Alt.* (Kandagatai, Sept. 14, 1876—Pot.; "Altai, Mongolie, Chaff."—Danguy, l.c.); *Dzhark.* (Kul'dzha vicinity, Aug. 12, 1878—Larionov).

IIIA. **Qinhai:** *Nanshan* ("Regio Tangut", 1880—Przew.); *Amdo* (Dzhakhar hills in Mudzhik river valley, July 6, 1890—Gr.-Grzh.).

IIIC. **Pamir** ("Eastern Pamir, east shore of Little Kara-Kul, 3720 m, July 16, 1894, Hedin"—Paulsen, l.c.).

General distribution: Mid. Asia (West. Pamir, wild growth), West. Sib. (south-east.), East. Sib. (south.), Far East, Nor. Mong. (Hent., Hang., Mong.-Daur., Fore Hing.), China (Dunbei, North, North-west, East, South-west), Korean peninsula.

Note. Paulsen (l.c.) placed the species recorded by him for East. Pamir under *A. tataricum* (= *A. odorum*) with a question mark. A.I. Vvedensky ("Flora Tadzhikistana" [Tajikistan Flora] 1963) reported *A. odorum* from West. Pamir (Bartang river basin) where this species is found in abandoned fields in cultivated valleys. As reported by Vvedensky, based on the views of P.A. Baranov and I.N. Raikova, *A. odorum* represents in Pamir a remnant crop formerly cultivated there (the species is grown in Asia as an edible and melliferous plant). Other species related to *A. odorum* to which the plant treated by Paulsen as *A. tataricum* could be assigned, do not grow in Pamir.

12. **A. oreoprasum** Schrenk in Bull. Sci. Acad. Sci. St.-Pétersb. 10, 23 (1842) 354; Rgl. in Izv. Obshch. lyubit. estestvozn., antrop., etnogr. 21, 2 (1876) 88; Rgl. in Acta Horti Petrop. 10, 1 (1887) 347; Deasy, In Tibet and Chin. Turk. (1901) 404; Danguy in Bull. Mus. nat. hist. natur. 17, 6 (1911) 5 and 20, 3 (1914) 140; Vved. in Fl. SSSR, 4 (1935) 162; Persson in Bot. Notiser, 4 (1938) 277; Fl. Kirgiz. 3 (1951) 62; Fl. Kazakhst. 2 (1958) 151; Fl. Tadzh. 2 (1963) 308; Ikonn. Opred. rast. Pamira [Key to Plants of Pamir] (1963) 87; Vved. in Opred. rast. Sr. Azii [Key to Plants of Mid. Asia], 2 (1971) 61. —**Ic.:** Rgl, l.c., 1876, tab. 14, figs. 7–9; Fl. Kirgiz. 3, Plate 12, fig. 1.

Described from Tien Shan (Kul'-Asu). Type in Leningrad. Plate III, fig. 1; map 3.

Rubble, rocky, melkozem (often loessial) slopes of mountains, steppe and desert clusters, pebble beds and sand; from foothills to upper mountain belt but predominantly in middle belt.

IB. **Kashgar:** *Nor.* (south. slope of Tien Shan, Uital river, on pebbles near brook, 1800–2100 m, July 2; Kara-Teke mountain range, on loessial hill slopes, 2100 m, June 7, 1889, Rob.; Vor dem Eingang zum Dschanart Tal, June 14–17; Kisyl-Sai, Plateauhöhe von Kara-Dschon, Aug. 2–3—1903, Merzb.; Uchturfan, June 2, 1908— Divn.; Bai town environs, Tukbel'chi village, 2530 m, No. 8371 (2700 m), No. 8387, Sept. 9, 1958—Lee and Chu); Muzart river valley, Chokarna area, 7–8 km beyond river exit from mountains, wormwood—forb arid steppe, 2080 m, No. 811, Sept. 7; same locality, Sazlik area, wormwood—feather-grass steppe, No. 888, Sept. 8; valley of Taushkan river [Kokshaal], 30 km beyond Uchturfan oasis, sympegma desert on rocky slope, 1700 m, No. 1061, Sept. 17—1958, Yun. and I.-F. Yuan'); *West.* (along road to Ulugchat from Kashgar, 15 km west of Kensu, on alluvium, 2500 m, No. 09672, June 1 and No. 09676, June 17; Sinkiang-Tibet highway 10 km north of Kalmangou, 2700 m, No. 00527, June 5; Upal environs, foothill flat, No. 00579, June 9 and No. 00587, Sept. 6—1959, Lee et al. (A.R. Lee); nor. slope of Kuen'-Lun', Akkëz-Daban pass, 110 km south of Kargalyk on Tibet highway, sympegma desert along debris cones, 2700 m, No. 347, June 5; same locality, 17 km south of Puss along Tibet highway from Kargalyk, sympegma desert along mountain trail of slope, No. 387, June 5; west. slope of Kuen'-Lun', foothill trails of nor. slope of King-Tau mountain range 5–8 km nor. of Kosh-Kulak settlement, natural sympegma desert on standing moraine, No. 549, June 9; nor. slope of Kuen'-Lun', Gëzdar'ya river valley 23 km nor. of Bulunkul' settlement, sympegma—wormwood grove, No. 582, June 11; Gëzdar'ya river valley, 40–42 km south of Upal oasis, sympegma desert along debris cones, No. 689, June 15; 58 km west-nor. west of Kashgar along road to Kensu spring, along ridges, Nos. 703, 741, June 17; Baikurt settlement 83 km nor.-west of Kashgar along road to Turugart, steppised desert along irrigation ditches, No. 850, June 19; 10 km south-east of Baikurt along old road to Kashgar from Turugart, saltwort-wormwood desert, along ravines between mud cones, 2300 m, No. 872, June 21—1959, Yun. and I.-F. Yuan'; 61 km west of Kashgar, 2170 m, No. 00373, June 17, 71 km west of Kashgar, 2250 m, No. 00379, June 17; in Baikurt region, on bank of irrigation ditch, 2200 m, No. 09703, June 19; 10 km south-east of Baikurt, in gorge, No. 09767, 2300 m, June 21—1959, Lee et al. (A.R. Lee (1959)); "Bostan-Terek, about 2400 m, July 10, 1924"—Persson, l.c.); *South.* (Keriya river valley, 2700 m, June 30; Keriya, mountain range along Nura river, on rocks, July 23, 1885—Przew.); *East.* (along Urumchi-Karashar highway, 1940 m, No. 6135, July 22 and No. 6143, July 23; Khetszin district, Logatou village, nor. of Bagrashkul' lake, 1750 m, No. 06258, Aug. 1—1958, Lee and Chu (A.R. Lee (1959)); 6–7 km south of Cherchen, on nor. slope in lower mountain belt, No. 9436, June 5, 1959—Lee et al. (A.R. Lee (1959)).

IIA. Junggar: *Tarb.* (east. trail of Saur mountain range on road to Burchum, mountain sheep's fescue steppe, July 4, 1959—Fedorovich); *Jung. Alatau* (Maili mountain range, Taidzhal hills, on edge of small dry gully, July 20, 1953—Mois.; Dzhair mountain range, 1–2 km nor. of Otu settlement along road to Toli and Chuguchak, steppe belt, along granite rocks, Aug. 4; south-west. outskirts of Maili mountain range, montane steppe belt, along rocky slope, Aug. 14; Shuvutin-Daba pass nor. of Sairam-Nur basin along road to Borotal, meadow steppe along mountain ridge, Aug. 18—1957, Yun., Li Shi-in, I.-F. Yaun'; Toli-Myaoergou, No. 2433, Aug. 4; vicinity Toli, steppe, Nos. 937, 969, 2479, Aug. 4; same locality, among rocks, 2020 m, No. 1769, Aug. 17; 10 km south of Bole [Dzhimpan'], 720 m, No. 2175, Aug. 29; 8 km nor. of Bole, No. 4760, Aug. 29—1957, Kuan); *Tien Shan* (Sairam-Nur, 2100 m, June 21, 1877—A. Reg.; Burkhantau, June 5; Urtas-Aksu, June 17—1878, Fet.; Yuldus, Taldy, 3000–3300 m, May 26; Cen. Dzhin along Tsagan-Usu, 1200–1800 m, June 7; Naryn-Gol on Tsagan-Usu, 1800–2400 m, June 10; foothills near Nilka, 1500–1800 m, June 21—1879, A. Reg.; Bogdoshan' mountain range, 1800 m, No. 5751, June 19; Khotun-Sumbul, 2550 m, No. 6264, Aug. 1; Bortu town environs, No. 6937, Aug. 2—1958, Lee and Chu (A.R. Lee (1959)); Ulyasutai-Chagan river valley, 4–5 km before Balinte settlement along road to Karashar from Urumchi, sympegma desert on rocky flank of valley, 1550 m, No. 151, Aug. 1; Khanga river valley 25 km nor.-west of

Balinte settlement along road to Karashar in Yuldus, standing moraine, steppe, Nos. 196, 198, 202, 209, 212, Aug. 1—1958, Yun. and I.-F. Yuan'; "Montagnes calcaires entre le Turkestan et la Mongolie, 1700 m, pres Gorgosse, July 17, 1895, Chaff."— Danguy, l.c. 1914; "Köl, hill-crevices, about 2700 m, Aug. 4, 1932"—Persson, l.c.); *Jung. Gobi* (west.: "Koustai, montagnes tres seches, July 25, 1895, Chaff."—Danguy, l.c. 1914).

IIIB. Tibet: *Chang Tang* (nor. foothills of Kuen'-Lun', Karasai area, on loess, 3000–3600 m, June 14; nor. slope of Russky mountain range [Keriya], Karasai village, clayey and sandy places, 3000 m, July 8—1890, Rob.; Polu, May 25, 1890—Grombch.; Polur, 1959—Lee et al. (A.R. Lee (1959)) [without precise date or number]; "Plateau near Polu, 3090 m, 1898"—Deasy, l.c.; Keriya river-basin, 4 km south of Polur settlement, mountain desert steppe, 3020 m, May 11; same locality, 2 km south-east of Polur settlement; wormwood desert on loessial sandy loams, May 11—1959, Yun. and I.-F. Yuan').

IIIC. Pamir (Ulug-Tuz gorge in Charlysh river basin, among rocks on river-bank, June 21, 1909—Divn.; "Jerzil, on dry clay slopes, about 3000 m, July 4, 1930"— Persson, l.c.).

General distribution: Jung.-Tarb., Nor. and Cent. Tien Shan, East. Pam.; Mid. Asia (Alay mountain range).

13. **A. polyrhizum** Turcz. ex Rgl. in Acta Horti Petrop. 3, 2 (1875) 162 and 10, 1 (1887) 339, p.p., excl. pl. tangut.; Forbes and Hemsley, Index Fl. Sin. 3 (1905) 124; Danguy in Bull. Mus. nat. hist. natur. 20 (1914) 140; Vved. in Fl. SSSR, 4 (1935) 172; Grub. Konsp. fl. MNR (1955) 94; Fl. Kazakhst. 2 (1958) 154; Grub. in Bot. mat. (Leningrad) 19 (1959) 536; Vved. in Opred. rast. Sr. Azii [Key to Plants of Mid. Asia], 2 (1971) 62. —*A. polyrhizum β. przewalskii* Rgl. l.c. (1875) 162 and 10, 1 (1887) 339. —Ic.: Rgl. l.c. (1887), tab. IV, fig. 1; Fl. SSSR, 4, Plate X, fig. 3, 3,a.

Described from East. Siberia (Dauria). Type in Leningrad. Plate I, fig. 6; map 1.

Desert steppes and deserts, rubble and rocky slopes of mountains and hills, mud cones, arid solonchaks, rocks, sometimes pebble beds; plains, foothills, lower and middle hill belts; forms coenosis in desert onion steppes.

IA. Mongolia: *Mong. Alt.* (Tsitsiriin-Gol, pebble bed, July 10, 1877—Pot.; Altyn-Khadasu cliff between Shara-Bulak spring and Kosheta brook, July 13; around Tsakhir-Bulak spring, July 18; mountain cliffs between Taishiri-Ol and Burkhan-Ol, July 18—1894, Klem.; before Dzasaktu-Khan camp, rubble slope, Aug. 7; Khara-Dzarga hills, Boro-Gol river valley, rocks, Aug. 24; Khara-Dzarga hills, Shutyn-Gol river valley, pebble bed, Aug. 28; foothills in east. part of Mong. Altay, near Borin-Khuduk well on rocky slopes, Sept. 3—1930, Pob.; right bank of Tsetseg-Gol 5 km south-west of Tsetseg-Nur lake, onion-feather-grass desert steppe, Aug. 12, 1945; pass through Dobtsig-Khuren-Nuru, steppe in hillocky area, Aug. 11, 1947—Yun.: east. extremity of Mong. Altay, large gorge along road to Bain-Undur, rubble slope, Aug. 25; Bain-Tsagan somon, east. extremity of Gichigin-Nuru mountain range, onion-feather-grass steppe, Aug. 27—1948, Grub.); *Cent. Khalkha* (Cent. Kerulen, steppe close to Bain-Erkhitu hill; same locality, near Dzhirgalantu pass; same locality, steppe near Shokhun hill—1899, Pal.; Sangin-Dalai, July 14 and 16, 1921—Lis.; Khukhu-Khoshun, rocky slopes, July 24, 1926—Lis.; south-east. Khangai foothill, Khai area, Aug. 13–16, 1926—Gus.; 375 miles from Urga, Aug. 1927—Terekhovko; on way from Ulan-Bator to Delger-Khangai, Shabalagin-Dzak cent. plateau, onion steppe, July 29, 1931—Ik.-Gal.; 5 km east of Choiren-Ul, Gashyun-Gol area, steppised derris grove on terrace above meadow, Aug. 23, 1940—Yun.; Erdeni-Dalai somon, 15 km south of intersection of old Dalan-Dzadagadsk road with caravan track running into Tsogtu-Chindamani somon, onion-feather-grass desert steppe, July 17; 1 km south of Erdeni-Dalai somon, onion-forb steppe, July 17; Idermeg somon, 5 km east of Idermet-Khid, in steppe, Aug. 30—1943, Yun.; 150 km south of Ulan-Bator, hillocky wormwood-feather-grass steppe, Aug. 11, 1950—Kal.); East. Mong. (Bain-Nor lake [south. extremity of Buir-Nor lake], sandy soil, June 19; Buir-Nor basin, Ikhtyr lake, July 6—1899, Pot. and Sold.; Manchuria railway station, solonchak nor. of station, July 26, 1925—Gordeev; Dariganga, around Tukhumyn-Gobi, Aug. 23, 25, and 26; Dariganga region, from Choiren to Naran area along Sair-Usin track, Dabastyn-Gobi, Sept. 5 and 8, 1927, Zam.; 70-75 km south-east of Choibalsan along road to Tamtsag, Erien-Obo area, onion-*Reaumuria* grove on solonetz soil, Aug. 3; Khalkha-Gol somon, 13–15 km nor.-nor.-east of Bain-Buridu along road to Khamar-Daba, Aug. 10; 50–55 km west-nor.-west of Khamar-Daba along road to Buir-Nur lake, vicinity of solonetz lowland, Aug. 11; south-east. fringe of Buir-Nur lake, 2–3 km from bank, wild rye-onion steppe, Aug. 14; 15–17 km nor.-east of Khara-Nur lake, wild rye meadow, Aug. 18—1949, Yun.; 65 km nor.-west of Matad somon along road to Choibalsan, meadow-feather-grass steppe, Aug. 15, 1949—Yun.; Tamtsag somon in Modon-Obo—Bulan-Khoolai region, derris grove on solonchak floor of valley, June 17; in Bayan-Ula region south-east of Khogon somon, Aug. 14, 1956—Dashnyam; Choibalsan somon, nor. of Enger-Shand, wild rye steppe, Aug. 4 and 5; same locality, 60 km nor. of Khabirsh border post, Aug. 12—1959, Dashnyam); Depr. Lakes (desert along Chon-Kharikha river, mountain range near Khara-Usu lake, Aug. 13 and 14, 1879—Pot.; between Dzabkhyn river and one of its tributaries, on arid solonchak, July 16, 1896—Klem.; Shargain-Gobi desert, on way from Borin-Khuduk well to Gol-Ikhe, rocky mountain trail, Sept. 4, 1930—Pob.; Daribi somon, 10–15 km south of Durge-Nur lake, onion-feather-grass desert steppe, Aug. 26—1944, Yun.; Tonkhil' somon, along road to Tonkhil' somon, 5 km from its emergence into gorge from Shargain-Gobi, steppe along mountain trail, Sept. 6, 1948—Grub.; left bank of Khungui along Dzapkhyn somon—Erdene-Khairkhan somon road, Altyn-Ula, montane steppe along rocky slope, Aug. 20; nor. mountain trail of Khasagtu-Khairkhan 3 km from Taishiri-Daribi pass, saltwort-onion desert, Aug. 22, 1972—Grub. et al.; "Envirens de Kobdo, Sept. 22, 1895, Chaff."—Danguy, l.c.); Val. Lakes (right bank of Ongiin river opposite Ongiin urton on arid solonchak, July 28, 1893—Klem.; near Dylger-Bulak spring, cliffs, June 26, 1894—Klem.; rocky slope to Tuin-Gol valley, Sept. 3; Tuin-Gol lowland, pebble bed, Sept. 6—1924, Pavl.; Naryn-Khara ridge, rubble-sand, Aug. 25, 1925—Glag.; Orok-Nur lake-basin, Naryn-Khara mountains, Aug. 12, 1926—Tug.; 5 km east of Tatsiin-Gol along road to Bain-Khongor from Arbai-Khere, gorge fringe, July 9, 1947—Yun.; Dzhinsatu somon, right bank of Tuiin river 15 km beyond somon, sand and sand mounds, Aug. 16, 1949—Kal.; Barun-Bayan-Ulan somon, Ulan-Toirim area, right bank of Tatsiin-Gol, among chee grass scrubs, July 25, 1951—Kal.; nor. of Serkhe-Ul, 50 km along road to west of Bu-Tsagan somon, smoothened low hillocky area, onion steppe, July 25, 1972—Grub. et al.); Gobi-Alt. (Khurkhu mountain range, Aug. 31, 1873—Przew.: south. slope of Tostu mountain range, broad valley floor, Aug. 18; valley between Tostu and Nemegetu mountain ranges, Aug. 19; nor. slope of Nemegetu mountain range, on rocks, Aug. 22—1886, Pot.; steppe near Dzhirgalante well, July 1924—

Gorbunova; Ikhe-Bogdo mountain range foothill, washed moraine, Aug. 24, 1926—Tug.; Bain-Tsagan mountain range, Shirikiin-Khid, Sept. 3, 1927—Simukova; Gurban-Saikhan mountain range, pass between Dzun-Saikhan and Dundu-Saikhan, steppe on flat ridge, July 22; Noyan somon, intermontane plain nor. of Noyan-Bogdo mountain range, onion desert steppe, July 24; 30–32 km west-south-west of Noyan somon along road to Obotu, smoothened hillocky area, onion-tansy steppe, July 26—1943, Yun.; Bain-Leg somon, Legiin-Gol valley 12 km west of camp in somon, Khalbagant-Ula mountain trail, saltwort desert, July 26; south-west. slope of Ikhe-Bogdo mountain range, Narin-Khurimt gorge, 2440–2500 m, July 30; Sever somon, Nemegetu-Nuru mountain range, nor. slope below rocks, 2700 m, Aug. 8; Tostu-Nuru mountain range, main peak of Sharga-Morite, on peak and west. slope, 2300–2565 m, Aug. 15; Bain-Undur somon, south. sloping trail to nor.-west of Dzadagat-Khere, feather-grass steppe, Aug. 25—1948, Grub.); *East. Gobi* (70 km before Choiren, steppe, on basalts, Aug. 24, 1926—Lis.; Khoshun-Ikhe-Dulan, Aug. 19, 1928—Lebedev; Dulan-Khoshun along road to Yaman-Ikhe, Aug. 20 and 21; before entering Udinsk lowland, Aug. 24; on plateau before descent into Tsongin-Gobi, near Sanchir well, Aug. 31; Argali mountain range and Khodatyin-Khuduk environs, Sept. 5—1928, Shastin; Kalgansk road between Sain-Usu well and Toskho-Nur lake, rubble semidesert, Aug. 15; same locality, around Baga-Ude well, along slopes and in valley of Balingote hill, Aug. 21; same locality, Dolochelut ridge, on loose sand, Aug. 23—1931, Pob.; road from Ulan-Bator to Gurban-Saikhan, Sept. 1, 1931—Krupenin; Dalan-Dzadagad town vicinity, feather-grass-desert steppe, July-Aug. 1939—Srmumazhab; Mandalin-Gobi somon, south. part of Mandalin-Gobi basin, saltwort-onion desert steppe, Aug. 14; 10 km nor.-east of Khara-Airik [Airag] somon along road to Choiren, Aug. 26; 60 km south-east of Choiren, saltwort-feather-grass desert steppe, Aug. 26; 16 km nor.-east of Sain-Shandy, tansy-feather-grass desert steppe, Sept. 2; Abdarantin-Tsab area south of Sain-Shandy, takyr (clay-surfaced) floor of lowland, Sept. 4; Tsogtu-Tsetsei somon, 75 km west of Khan-Bogdo along road to Dalan-Dzadagad, onion-feather-grass steppe, Sept. 2—1940, Yun.; 45 km west of Khan-Bogdo somon along road to Dalan-Dzadagad, feather-grass-onion steppised desert, 1943—Yun. [undated]; Dalan-Dzadagad, adventitious in kitchen gardens, July 21, 1943—Yun.; 30–40 km south-west of Undur-Shilch somon, Tash area, rather low slopes of knolls, July 27, 1946—Yun.; Mandal-Obo somon, Bain-Dzak-Ula area, eroded red sandstones, plateau and upper part of slopes, Oct. 21, 1947—Grub. and Kal.; Kholtu somon, plain 6 km nor.-east of Oldakhu-Khid monastery, onion-feather-grass steppe, Aug. 15, 1950—Kal.; Bailinmyao, desert steppe, 1959—Ivan.; 20–25 km south of Delger somon, along road to Bain-Dzak, feather-grass-onion steppe, Aug. 12; 43 km nor.-east of Dalan-Dzadagad, wormwood-snake weed-onion gully, Aug. 14—1962, Lavrenko, Yun., Dashnyam); *West. Gobi* (Tsel' somon, submontane flat south of Mongol. Altay, Tszakhoi-Tszaram, desert steppe, Aug. 20, 1943—Tsebigmid); *Alash. Gobi* ([road from Dyn'yuan'in (Bayan-Khoto) to Gun'-Khuduk well], July 26, 1873—Przew.; Alashan mountains, 1880—Przew.; 30–32 km west-south-west of Noyan somon along road to Obotu, smoothened hillocky area, onion-tansy steppe, July 26, 1943—Yun.; 50 km on highway from Inchuan to Bayan-Khoto, rocky montane slopes, 1800 m, June 10, 1958—Petr.); *Khesi* (25–30 km north-west of Lan'chzhou town along road to Uvei, along loessial cliffs, Oct. 9, 1957—Yun. et al.; 45 km west of Yunchan town, nor. hilly Nanshan foothills, July 10; 60 km south-east of Chzhan'e town, high Nanshan foothills, 2200 m, July 12; 60 km west of Tszyutsyuan' town, pebble-loam submontane flat, July 28; Tszyutsyuan' town environs, nor. submontane pebble flat of Nanshan, Aug. 7; 55 km east of Chzhan'e, hillocky Nanshan foothills, Aug. 10—1958, Petr.).

IIA. **Junggar:** *Tarb.* (Semiz-Chii area, Aug. 8, 1876—Pot.; nor. trail of Saur mountain range, 13 km west-nor.-west of Kheisangou settlement along road from Burchum to Karamai, desert steppe, No. 1672, July 10, 1959—Yun. and I.-F. Yuan'); *Jung.* **Alatau** (Dzhair mountain range, road from Karamai to Chuguchak, Aug. 5, 1951—Mois.; Taidzhala mountains [Maili mountain range], on fringe of small arid ravine, 1200 m, July 20, 1953—Mois.; Dzhair mountain range, Otu picket, steppe, No. 867, Aug. 3; same locality, No. 936, Aug. 4—1957, Kuan; Dzhair mountain range, along road from Aktam to Otu and Chuguchak, along rubble slopes of knolls in desert-steppe belt, Aug. 3; same locality, 1 km south of Otu settlement along old road to Shikho from Chuguchak, feather-grass steppe with wormwood and onion, Aug. 9—1957, Yun. et al; 30 km nor. of Sairam-Nur lake, wormwood-onion semidesert on intermontane plain, 1150 m, Aug. 31, 1959—Petr.; "Intermontane valley on Chipeidzy-Chuguchak road, 600 m, Aug. 5, 1951, Mois."—Grub. l.c.); *Jung.* **Gobi** (nor., sand along south-east. bank of Ulyungur lake, Aug. 11 and 15, 1876—Pot.; Dzhungar basin, left bank of Urungu river near Dinsyan' settlement; *Anabasis salsa*-winter fat desert on rubble loam, July 13, 1959—Yun.; 60 km south of Ertai settlement, hillocky desert, July 16, 1959—Yun. and I.-F. Yuan'; west.: east of San'tai, in desert along road, No. 4740, Aug. 29, 1957—Kuan; south-west. part of Ebi-Nur basin, trails of Kyzemchek mountain range, descending from Santai into Utai on Sairam-Tszin'kho highway, desert-steppe belt, onion-desert steppe, No. 1751, Sept. 1, 1957—Yun. et al.; "Steppes entre l'Ebi-Nor et l'Irtich, 1400 m, Aug. 3, 1895, Chaff."—Danguy, l.c.; south.: 517 km nor.-east of Sairam-Nur lake on highway, No. 2199, Aug. 28, 1957—Kuan; east.; near Barkul' lake, in desert, No. 4497, Sept. 28, 1957—Kuan).

General distribution: Jung.-Tarb. (Tarbagatai); East. Sib. (Dauria: south.); Nor. Mong. (Hang.: south.), China (Dunbei).

Note. Regel (Rgl. l.c. 1875) described from Alashan var. *przewalskii* and in his first description distinguished it from typical individuals of this species by fibrous-split bulb tunic very closely surrounding stem. Later (l.c. 1887), while ascribing this variety for Tarbagatai, Cen. Khalkha and Depr. Lakes, Regel additionally characterised it with filaments of inner stamens greatly enlarged (for 1/2 length or slightly more) and with more obtuse teeth. A.I. Vvedensky (l.c. 1935) points out that var. *przewalskii* bears much larger (6 mm long) flowers and suggests that this variety merits further investigation. N.V. Pavlov and P.P. Polyakov in "Flore Kazakhstana" [Flora of Kazakhstan] placed it among synonyms of *A. polyrhizum* without comment.

A study of vast herbarium material (including type specimens) of *A. polyrhizum* revealed no justification for recognising it as a distinct variety. The type specimen of var. *przewalskii* does not differ from the type variety in size of perianth lobes; nor is there any difference in length of enlarged part of filaments of inner stamens. Teeth of filaments vary in *A. polyrhizum* throughout its tribution range from more acute and long to very short and subobtuse; length of perianth lobes varies from 4 to 5 (6) mm. *A. subangulatum* Rgl., closely related to *A. polyrhizum*, possesses larger perianth lobes (see below).

The Herbarium of the Romaro Botanical Institute has a single specimen of *A. polyrhizum* which is entirely typical. As stated on the label, it was grown in the Botanical Garden from bulbs collected by N.M. Przewalsky in Qinghai (regio Tangut), but *A. polyrhizum* does not grow in that region. In all probability, the original label was mislaid.

14. **A. przewalskianum** Rgl. in Acta Horti Petrop. 3, 2 (1875) 164 and 10, 1 (1887) 343; Kanitz in Széchenyi, Keletázsiai ut jának 2 (1891) 59; id. in Széchenyi, Wissensch. Ergebn. 2 (1898) 734; Diels in Futterer, Durch

Asien (1903) 7; id. in Filchner, Wissensch. Ergebn. 10, 2 (1908) 248; Danguy in Bull. Mus. nat. hist. natur. 17, 6 (1911) 5; Hao in Bot. Jahrb. 68 (1938) 587. —Ic.: Rgl. in Acta Horti Petrop. 10, 1, tab. IV, fig. 2.

Described from Qinghai (Nanshan). Type in Leningrad. Plate III, fig. 5; map 3.

Meadows, river-banks, rocks; alpine and upper part of middle mountain belt.

IA. Mongolia: *Khesi* (between Yunchen'syan' and Gan'chzhou towns, July 19, 1875—Pias.).

IIIA. Qinghai: *Nanshan* (South Tetung mountain range, in alpine meadow, July 31, 1872—Przew., typus!; same locality, on sandy bank of Tetung river, 2150 m, Aug. 8, 1880—Przew.; nor. slope of Humboldt mountain range, Ulan area, arid meadow, 3600 m, July 6; South Kukunor mountain range, Bain-Gol valley, clayey soil, 3600 m, July 26—1894, Rob.; Kukunor lake, 1908—Czet.; Tetung river valley [Tetungkhe], near stud farm, along slopes of knolls, grass-forb steppe, 2800 m, Aug. 20, 1958—Dolgushin; meadow along east. bank of Kukunor lake, 3210 m, Aug. 5, 1959—Petr.; "Hoang-ho Schluchten oberhalb Balekun-gomi, Sept. 27 [undated] Futterer"—Diels, l.c. 1903; "Sining-fu, 1904, Filchner"—Diels, l.c. 1908; "Sining-Fou, 2400 m, July 12, 1908, Vaillant"—Danguy, l.c.; "Kokonor, Schalakutu, 3400 m, Aug. 18, 1930"—Hao, l.c.); *Amdo* ("Kuncsang-fu [Kutschang-fu], Aug. 23, 1879, Széchenyi"—Kanitz, l.c.).

IIIB. Tibet: *Weitzan* (Burkhan-Budda mountain range, mid-portion, 3450–3900 m, swampy grasslands, Aug. 14, 1884—Przew.; nor. slope of Burkhan-Budda mountain range, Khatu gorge, rock crevices, 3300–3900 m, July 11, 1901—Lad.).

General distribution: China (North-West: Gansu, South-West: Sichuan).

Note. The species varies in perianth size—(3) 4–5 (5.5) mm long—pedicel length (usually 2–2.5 times, rarely slightly longer than perianth) as well as in length of beak of spathe—from 2 (2.5) to 0.7 (0.5) cm. Specimens collected from Tetung river basin of Qinghai possess the longest spathe beaks. *A. przewalskianum* is closely related to *A. stoliczkii* Rgl. (see below) as well as to *A. henryi* described from Cen. China (Hubei) by C.H. Wright in Kew Bull. (1895) 119. Wright differentiates it from *A. przewalskianum* in very long leaves and pedicels, shorter and broader spathe and fewer flowers in inflorescence.

15. **A. schrenkii** Rgl. in Acta Horti Petrop. 3, 2 (1875) 172.—*A. bogdoicolum* Rgl. in Acta Horti Petrop. 6 (1880) 530; Vved. in Fl. SSSR, 4 (1935) 151; Fl. Kazakhst. 2 (1953) 145; Vved. in Opred. rast. Sr. Azii [Key to Plants of Mid. Asia], 2 (1971) 56. —*A. strictum* auct. non Schrad.: Vved. in Fl. SSSR, 4 (1935) 151, p.p., quoad syn. *A. schrenkii*; Fl. Kazakhst. 2 (1958) 145, p.p., quoad syn. *A. schrenkii*; Vved. in Opred. rast. Sr. Azii [Key to Plants of Mid. Asia], 2 (1971) 56, p.p., quoad syn. *A. schrenkii*.

Described from East. Kazakhstan (Fore Balkhash, Alakul' lake environs). Type in Leningrad.

Alpine meadows.

IIA. Junggar: *Tien Shan* (Bogdo mountain, 3000 m, July 25, 1878—A. Reg.).
General distribution: Jung.-Tarb. (Jung. Alatau.).

Note. In the works cited, the species was treated as a synonym for *A. strictum* Schrad. Concomitantly, *A. schrenkii* is identical to *A. bogdoicolum* Rgl. described from Junggar (specimen cited above is type of *A. bogdoicolum* Rgl.) Types of both species are entirely identical. *A. schrenkii* is very closely related to *A. amphibolum* as well as to *A. strictum*. The affinities of these species call for further investigation (see also note under *A. amphibolum*).

16. **A. splendens** Willd. ex Schult. et Schult. f. in Roem. and Schult. Syst. Veg. 7, 2 (1830) 1023; Franch. Pl. David. 1 (1884) 304; Vved. in Fl. SSSR, 4 (1935) 150; Kitag. Lin. Fl. Mansh. (1939) 132. —*A. lineare* auct. non L.: Rgl. in Acta Horti Petrop. 10, 1 (1887) 344, p.p.

Described from Siberia. Type in Berlin (B)?

IA. **Mongolia:** *East. Mong.* ("Bois frais des hautes montagnes de l'Ourato, July 1866, David"—Franch, l.c.).

General distribution: East. Sib. (south.), Far East, China (Dunbei), Japan.

17. **A. stoliczkii** Rgl. in Acta Horti Petrop. 3, 2 (1875) 160.

Described from Himalayas (Kashmir). Type in Leningrad.

Moist river-banks and pebble beds, alpine meadows, gullies, moraines; alpine, rarely middle hill belt.

IB. **Kashgar:** *West.* (upper Tiznaf river 15 km beyond Kyude settlement on road to Saryk-Daban pass, short-grass meadow along brook, 3400 m, No. 233, June 1; 2–3 km south-west of Akkëz-Daban pass, 120 km from Kargalyk along Tibet highway, montane sympegma desert, along gully with limestone outcrops, No. 353, June 5—1959, Yun. and I.-F. Yuan').

IIA. **Junggar:** *Tien Shan* (south. slope of Tien Shan, 25 km nor.-west of Balinte settlement and above along Khanga river valley, along road to Yuldus from Karashar, on pebble beds of floodplain and in meadows, No. 189, Aug. 1; same locality, standing moraine along gullies, No. 201—1958, Yun. and I.-F. Yuan'; Khetszin district, Logotou village, nor. of Bagrashkul' lake, 2550 m, No. 06282, Aug. 1, 1958—Lee and Chu (A.R. Lee (1959)).

IIIB. **Tibet:** *Chang Tang* (Keriya mountain range, along Kyzylsu river, 3150 m, July 14, 1885—Przew.); *Weitzan* (Yantszytszyan river basin, Donra area on Khichu river, on rocks, July 16, 1900—Lad.).

IIIC. **Pamir:** (Ulug-Tuz near its confluence with Charlysh river, on rock placers, July 1, 1909—Divn.; upper Kashkasu river, below moraine, about 3500 m, July 5; upper Lanet river, moraine, July 20; Kara-Dzhilga river, alpine pasture, 4000–5000 m, July 22; Taspetlyk area, 4000–5000 m, July 25; environs of Takhtakorum pass, in upper courses of Kulan-Aryk and Takhta rivers, 4500–5500 m, Aug. 1; Takhta river gorge, 2700 m, Aug. 2—1942, Serp.).

General distribution: Himalayas (Kashmir), China (South-West.—Sichuan).

Note. Closely related to *A. przewalskianum* Rgl. (see above), from which it differs in very short (1–2 mm) barely visible beak of spathe as well as very short pedicels, as long as, or shorter than flowers, rarely somewhat longer. Regel (Rgl. l.c. in key) differentiates *A. stoliczkii* from *A. przewalskianum* in that the enlarged part of filaments of inner stamens is as long as unenlarged subulate part in the former and not shorter as in *A. przewalskianum* and other species. Actually, however, the enlarged and unenlarged parts of filaments are equal in both species.

18. **A. strictum** Schrad. Hort. Goett. (1809) 7; Rgl. in Acta Horti Petrop. 3, 2 (1875) 164; Vved. in Fl. SSSR, 4 (1935) 151; Fl. Kazakhst. 2 (1958) 145; Vved. in Opred. rast. Sr. Azii [Key to Plants of Mid. Asia], 2 (1971) 56.

—*A. lineare* var. *strictum* (Schrad.) Kryl. Fl. Zap. Sib. 3 (1929) 626. —*A. lineare* auct. non L.: Rgl. in Acta Horti Petrop. 10, 1 (1887) 344, p.p.; Grub. Konsp. fl. MNR (1955) 93, p.p., quoad pl. ex Alt. mong. et gobic. —Ic.: Schrad. l.c. tab. 1.

Described from Siberia. Type in Göttingen?

Rocky slopes of mountains and knolls, montane steppes, larch forests and scrubs; from foothills to middle mountain belt.

IA. **Mongolia:** *Mong. Alt.* (Taishiri-Ula, in forest, July 15, 1877—Pot.; same locality, in forest, July 16, 1894—Klem.; 8 km south-east of Dzasaktu-Khan camp, larch forest along mountain slope, Aug. 9; Khara-Dzarga mountain range, Sakhirsala river valley, western rocky slopes, Aug. 22; nor. slope of Khara-Dzarga mountain range, Khairkhan-Duru environs, larch forest, Aug. 25—1930, Pob.; 10 km south-east of Yusun-Bulak, central part of nor. trail of Khan-Taishiri mountain range, July 14; Tamchi somon, south. slope of Tamchi-Daba mountain range, montane steppe, July 16; Khubchin-Nuru mountain range west of Adzhi-Bogdo, along slopes of knolls, Aug. 3, 1947—Yun.); *Depr. Lakes* (Ubsa lake environs, Ulan-Natsyn river valley, July 16; southern tributaries of Kharkhira river, July 24—1879, Pot.; south. slope of Khan-Khukhei mountain range, scrubs, July 22, 1945—Yun.), *Gobi-Alt.* (Dundu-Saikhan hills, middle belt of west. slopes, July 9, 1909—Czet.; Artsa Bogdo, canyons and moist slopes, 2010–2250 m, 1925—Chaney).

IIA. **Junggar:** *Tarb.* (Dachen [Chuguchak] town vicinity, No. 1675, Aug. 14; nor. of Dachen, on slope, No. 2913, Aug. 17—1957, Kuan); *Jung. Alatau* (upper Borotala, 1800 m, Aug.; upper Khorgos, Aug.—1878, A. Reg.; mountains in Toli district, on slope, No. 2604, Aug. 7, 1957—Kuan; south-west. fringe of Maili mountain range 10–12 km from Karaganda pass along road to Junggar exit, along rocky slope in steppe belt, Aug. 14, 1957—Yun. et al.); *Tien Shan* (southern bank of Sairam lake, July 23, 1877—A. Reg.; Bogdo mountain, 2400–2700 m, July 1878—A. Reg.; hill south of Urumchi, No. 343, July 14, 1956—Ching; Kuitun river basin, Bain-Gol creek right bank, south of Tushandza settlement, montane steppe in lower part of forest belt, June 29, 1957—Yun. et al.; Bulun'tai nor. of Bagrashkul' lake, No. 1170, Aug. 7, 1957—Kuan); *Jung. Gobi* (west.: Savan district, 6 km south-west of Shichan, on slope, 1700 m, No. 2294, July 26, 1957—Kuan).

General distribution: Jung.-Tarb., Nor. Tien Shan; Europe (Cent. Europe and European USSR), West. Sib. (south.), East. Sib. (south.), Far East, Nor. Mong. (Hent., Hang.).

19. A. subangulatum Rgl. in Acta Horti Petrop. 10, 1 (1887) 340; Forbes and Hemsley, Index Fl. Sin. 3 (1905) 125. —*A. polyrhizum* auct. non Rgl.: Rehder in J. Arn. Arb. 14 (1933) 5. —Ic.: Rgl. l.c. tab. V, fig. 1, 1,a, l,b,.

Described from Qinghai. Type in Leningrad. Map 2.

Sandy-pebble lake-banks, sandy steppes, clayey slopes, upper mountain belt.

IA. **Mongolia:** *Alash. Gobi* (south. Alashan [from Gansu boundary in Alashan to Tsokhor-Tologoi well], on sand, Aug. 25, 1880—Przew.).

IIIA. **Qinghai:** *Nanshan* (on bank of Kukunor lake, July 14; Kukunor lake, on sandy-pebbly banks, July 15, 1880, Przew., typus!; Sharagol'dzhin river valley, Paidza-Tologoi area, sandy-pebble steppe, 3000 m, July 11, 1894—Rob.; nor. slope of Humboldt mountain range near pass, Chansai-Lontszy area, along loessial slopes, July 22, 1895—Rob.; Gunkheisk basin 20 km west of Gunkhe, steppe, 2980 m, Aug. 6, 1959—Petr.); *Amdo* (Mudzhik river midcourse, Dzhakhar mountains, July 7, 1890—

Gr.-Grzh.; nor. slope of Burkhan-Budda mountain range, Khatu gorge, clayey slopes and pebble beds, 3150 m, June 24, 1901—Lad.; "Eastern Tibet, Ba Valley, Rock"— Rehder, l.c.).

General distribution: China (North-West: Gansu).

Note. Closely related to *A. polyrhizum*, from which it differs in longer, subacute perianth lobes narrowing upward and filaments of inner stamens enlarged for 2/3 or 3/4 length.

20. **A. teretifolium** Rgl. in Acta Horti Petrop. 5 (1878) 629 and 10, 1 (1887) 325; Vved. in Fl. SSSR, 4 (1935) 186; Fl. Kirgiz. 3 (1951) 74; Fl. Kazakhst. 2 (1958) 166; Vved. in Opred. rast. Sr. Azii [Key to Plants of Mid. Asia], 2 (1971) 67. —*A. grimmii* Rgl. in Trautv., Rgl. Maxim. et Wincl. Dec. pl. nov. (1882) 10. —*A. tekesicolum* Rgl. l.c. 10, 1 (1887) 350. —*A. deserticolum* M. Pop. in Bot. Mat. (Leningrad) 8 (1940) 75; Fl. Kazakhst. 2 (1958) 173.

Described from specimens grown from bulbs collected in Junggar Alatau (Altyn-Emel' mountain range). Type in Leningrad. Plate IV, fig. 2.

Rubble and rocky slopes of mountains; from foothills to upper mountain belt.

IIA. Junggar: *Tien Shan* (along Tekes river, 2500 m, July 7, 1871—Przew.; bank of Tekes river, 1350–1500 m, Aug. 1877; Kyzemchek mountains near Sairam lake, 2100–2700 m, July 29, 1878; Kash river between Ulastai and Nilka, 900–1200 m, June 30; Kash river, 900–1200 m, July 3; Dzhirumtai, 900–1200 m, July 2; Mongoto, 3000–3300 m, July 4; same locality, 2700 m, Aug. 2—1879, A. Reg.; along road between Ili and Dzhagastai, on slope, No. 3156, Aug. 7; 26 km south of Chzhaosu along road to Tekes, No. 991, Aug. 17—1957, Kuan; Ili river valley, Ketmen' mountain range, 1 km before Sarbushin settlement on way from Kzyl-Kure to Ili, along rocky slope in steppe belt, Aug. 21, 1957—Yun.); *Jung. Gobi* (south.: nor. of Urumchi, 1700 m, Nos. 1218, 1229, Aug. 26, 1957—Kuan).

General distribution: Jung.-Tarb., Nor. and Cent. Tien Shan.

Note. This species exhibits extremely close affinity with *A. inconspicuum* Vved. and *A. barszewskii* Lipsky but not with *A. korolkowii* Rgl., *A. weschnjakowii* Rgl. and *A. globosum* M.B. ex Redouté, together with which Vvedensky placed it in the work cited. It differs from these species (apart from other characteristics) in reticulately fibrous bulb tunic for which reason it should be placed in a group of species bearing this extremely significant characteristic. It differs from *A. inconspicuum* and *A. barszewskii* in long beak of spathe (8–10 mm, not 2–3 mm) and extent of adnation of filaments of stamens with perianth: in *A. teretifolium*, enlarged parts of filaments are connate and adnate with perianth for 1/3 (or less) of length, while in the other species mentioned, this adnation is 1/2 or 2/3.

Section Rhizirideum

21. **A. angulosum** L. Sp. pl. (1753) 299; Rgl. in Acta Horti Petrop. 3, 2 (1875) 143; Kryl. Fl. Zap. Sib. 3 (1929) 619; Gr.-Grzh. Zap. Mong. 3, 2 (1930) 807; Vved. in Fl. SSSR, 4 (1935) 164; Fl. Kazakhst. 2 (1958) 152; Vved. in Opred. rast. Sr. Azii [Key to Plants of Mid. Asia], 2 (1971) 61.

Described from Siberia. Type in London (Linn.).

Alpine meadows.

IIA. **Junggar:** *Cis-Alt.* ("Valley of Kran river upper course, alpine meadow, July 6, 1903"—Gr.-Grzh. l.c.).

General distribution: Fore Balkh. (? Zaisan basin); Europe (including European part of the USSR), West. Sib., East. Sib. (south-west).

22. **A. anisopodum** Ledeb. Fl. Ross. 4, 1 (1852) 183; Trautv. in Acta Horti Petrop. 1, 2 (1872) 193. Forbes and Hemsley, Index Fl. Sin. 3(1905) 125; Kryl. Fl. Zap. Sib. 3 (1929) 622; Vved. in Fl. SSSR, 4 (1935) 174; Kitag. Lin. Fl. Mansh. (1939) 130; Grub. Konsp. fl. MNR (1955) 92; Fl. Kazakhst. 2 (1958) 156; Vved. in Opred. rast. Sr. Azii [Key to Plants of Mid. Asia], 2 (1971) 62. —*A. tenuissimum β. anisopodum* (Ledeb.) Rgl. in Acta Horti Petrop. 3, 2 (1875) 157 and 10, 1 (1887) 314.

Described from East. Siberia. Type in Leningrad. Plate IV, fig. 8.

Arid and desert steppes, deserts, rocky and rubble steppe slopes of mountains and knolls, rocks, sand, sandy banks of rivers and lakes and pebble beds; in plains, foothills and lower mountain belt.

IA. **Mongolia:** *Khobd.* (Iter ad Kobdo, June 20, 1870—Kalning; Tszusylan, on pebble bed, June 29, 1877—Pot.; foothills nor.-east of Bairimen-Daban slope, June 20; Uryuk-Nor lake, on rocks, June 21; desert near Ulan-Daba, June 23, 1879—Pot.; Sagli river valley, Aug. 6; Achit-Nor lake near Beku-Muren, around granite mountains, Sept. 13—1931, Bar.); *Mong. Alt.* (Urten-Gol river, on sand, June 30; Dolon-Nor, pebble bed, July 8—1877, Pot; Tsagan-Gol upper valley, moraines between Kharsala and Prokhodna, June 30, 1905—Sap.; steppe near Kobdo, July 18, 1906—Sap.; Khara-Dzarga mountains, Sakhir-Sala river valley, southern rubble slopes, Aug. 23, 1930—Pob.; south-west. slope of Taishiri-Ul near Mogyin-Baishing, feather-grass steppe, July 12; Tamchi somon, nor. trail of Bus-Khairkhan mountain range, gully, July 17—1947, Yun.); *Cen. Khalkha, East. Mong., Depr. Lakes* (desert on bank of Kharkhira river, July 9; desert between Kobyuden-Khuduk well and Deeren-Nor lake, on sand, Aug. 4—1879, Pot.; 10 versts (1 verst = 1.067 km) from Ubsa lake, July 3, 1892—Kryl.; between Dzapkhyn river and Bulyn-Bulak spring, in steppe, June 29, 1894—Klem.; Santa-Margats somon nor. of Margats-Ul, closer to Tsagan-Nur lake, snakewort-feather-grass steppe, Aug. 18, 1944; Dzun-Gobi somon, Borig-Del' sand 6–7 km east of Baga-Nur lake, hummocky sand, July 25, 1945; 3 km south of Ulangom on road to Kobdo, along rocky slope of granite knoll, July 27, 1945—Yun.); *Val. Lakes* (along road to Kuku-Khoto from Kobdo, descent to Baidarik river, on cliff, June 19, 1894—Klem.); *Gobi Alt.* (Gurbun Saikhan, dry canyon walls, 2010 m, 1925—Chaney; Ikhe-Bogdo mountain range, Bityuten-Ama, hill slopes, Aug. 12, 1927—Simukova; pass between Dzun- and Dundu-Saikhan hills, lower belt, forb steppe on floor of arid gully, July 22; foothill plain nor. of Gurban-Saikhan mountain range, snakewort-feather-grass desert steppe, July 22; 2 km south-west of Noyan somon, feather-grass desert steppe in hillocky area, July 25; 10–12 km west of Noyan somon, intermontane valley, feather-grass desert steppe, July 26; west. extremity of Baga-Bogdo mountain range, upper part of mountain trail, wormwood-forb desert steppe, Sept. 17—1943, Yun.); *East. Gobi* (granite mountain 25 km south of Ude, July 19, 1928—L. Shastin; Khan-Bogdo somon, Khoir-Ultzeitu area near Khutuk-Ul, desert steppe, June–Aug. 1930—E. Kuznetsov; Bain-Dzak area, pebble bed steppe, Sept. 2, 1931—Krupenin; Sain-Tsagan somon, 10–15 km south of Mandal-Gobi on old Ulan-Bator–Dalan-Dzadagad road, desert steppe along hillocky ridged valley, Aug. 19; Khara-Airik somon, elevated plain 20–30 km south-east of Buterin-Obo along Ulan-Bator–Sain-Shanda highway, feather-grass—wormwood desert steppe, Aug. 27—1940, Yun.; Lus somon, Khatu-Tugrik area, intermontane basin near lake, sandy coastal strip, July 14, 1950—Kal.; Bailinmyao town, desert steppe, 1959—Ivan.); *Alash. Gobi* (Alashan mountain range, July 7, 1873—Przew.; Alashan mountain range, Tszosto gorge, sandy soil in lower belt, June 18, 1908—Czet.; "Ho Lan Shan, 1923, Ching"—Walker, l.c.); *Ordos* (Dzhungor camp in Ordos, clayey-sandy soil, Aug. 13; Baga-Edzhin-Khoro, Aug. 18—1884, Pot.; 40 km south of Dalat town, Khantaichuan river valley, precipices of bedrocks, Aug. 4; 70 km south of Khangin town, red sandstones on ridge peak, Aug. 5; 30 km south of Dalat town, Inken-Obo outlier, Aug. 10; same locality, meadow in Huang He river valley, around Chzhandanchzhao village, Aug. 10—1957, Petr.).

IIA. **Junggar:** *Tarb.* (east. trails of Saur mountain range, 35 km nor. of Kosh-Tologoi settlement along road to Karamai in Altay, sheep's fescue-snakewort steppe, July 4, 1957—Yun. and I.-F. Yuan'} *Jung. Gobi* (nor.: south-east. sand bank of Ulyungur lake, Aug. 15, 1876—Pot.); *Zaisan* (left bank of Ch. Irtysh west of Cherektas town, hummocky sand, June 10, 1914—Schischk.; Zimunai, 1300 m, No. 10577, June 26; along road to Burchum from Khabakhe, on sand, July 18—1959, Lee et al. (A.R. Lee (1959)).

IIIA. **Qinghai:** *Amdo* (Huang He river upper course, May 1880—Przew.).

General distribution: Jung.-Tarb. (Tarbagatai); West. Sib. (Kazakhstan—east. hillocky terrain and Altay), East. Sib. (south.), Far East (south.), Nor. Mong., China (Dunbei, North, North-west), Korean peninsula, Japan.

Note. See note under *A. tenuissimum* L.

23. **A. bidentatum** Fisch. ex Prokh. in Mater. komiss. po issled. Mong. i Tuv. 2 (1929) 83, in adnot.; Prokh. in Izv. Glavn. bot. sada SSSR, 29, 5–6 (1930) 564; Vved. in Fl. SSSR, 4 (1935) 172; Kitag. Lin. Fl. Mansh. (1939) 130; Gordeev and Jernakov in Acta pedol. sin. 2 (1954) 278; Grub. Konsp. fl. MNR (1955) 92; Fl. Kazakhst. 2 (1958) 155; Vved. in Opred. rast. Sr. Azii [Key to Plants of Mid. Asia], 2 (1971) 62. —**Ic.:** Prokh. l.c. Plate 5; Fl. Kazakhst. 2, Plate 13, fig. 5.

Described from East. Siberia (Transbaikal). Type in Leningrad, Plate I, fig. 2.

Arid steppes, steppe melkozem, rubble and rocky slopes of mountains and knolls, rocks; plains and lower mountain belt.

IA. **Mongolia:** *Cen. Khalkha* (right bank of Ongiin river, plateau among fine rubble, July 27, 1893—Klem.; along Khara-Gol river, 900–1350 m, July 24–27, 1906—Novitski; Tsagan-Nor lake environs, on steppe, Aug. 5, 1925—Glag.; 60–80 versts south-west of Urgi [Ulan-Bator], valley of Sosyn-Khuduk river, dry bed, 1925; Ikhe-Tukhum-Nor lake environs, Khubinus-Michigun pass, June; same locality, steppe between nor. mountains and Ikhe-Tukhum-Nor, July 27—1926, Zam.; south-east Hangay foothills, Kholt area, July 13–16, 1926—Gus.; 50 km south-east of Ulangur workshop on highway from Ulan-Bator to Choiren—Sain-Shandu, feather-grass-wild rye steppe, July 23, 1941—Yun.; Undur-Untsa somon, 30 km nor.-nor.-west of somon along road to Arbai-Khere, feather-grass-wheat grass-wormwood steppe, July 6, 1941—Tsatsenkin; Tsinkir-Mandal somon, Tsinkiriin-Gol valley near Tsinkir-Dugang, steppe along rocky slope, July 23; Idermeg somon, 10 km east of Idermeg-Khida along road, feather-grass-snakewort steppe, Aug. 30—1949, Yun.; Unzhul somon, 35 km south-

south-west of Sorgol-Khairkhan-Ul, steppe along hillocky territory, Sept. 4, 1950, Lavr. et al.); *East. Mong., Val. Lakes* (right bank of Tatsiin-Gol river, July 20, 1893—Klem.; along road from Arbai-Khere, Garadain-Gol area, steppe, July 9, 1947—Yun.); *Ordos* (valley of Shukhen-Gol river, Aug. 20, 1884—Pot.; 50 km west of Dunshen town, arid steppe on top of knoll, Aug. 7, 1957—Petr.).

IIA. Junggar: *Jung. Gobi* (east.: south. bank of Barkul' lake, in intermontane depression, No. 2237, Sept. 28, 1957—Kuan).

General distribution: East. Sib. (south.), Nor. Mong. (Hent., Hang., Mong.-Daur., Fore Hing.), China (Dunbei).

Note. Plants of this species are sometimes found with filaments of stamens without or with barely developed teeth.

24. **A. dentigerum** Prokh. in Izv. Glavn. bot. sada SSSR, 29, 5–6 (1930) 563. —Ic.: l.c. fig. 4.

Described from Qinghai (Nanshan). Type in Leningrad.

River-banks in lower part of forest belt.

IIIA. Qinghai: *Nanshan* (South Tetung mountain range, on sandy bank of Tetung river in lower part of forest belt, 2250 m, Aug. 8, 1880—Przew., typus! along Itel'-Gol river, April 1885—Pot.).

General distribution: endemic.

Note. While describing this species, Ya.I. Prokhanov observed its similarity with *A. bidentatum* Fisch. ex Prokh. but, in combination of characters (without specifying which), declared it differed distinctly from the latter. *A. dentigerum* is very closely related to *A. bidentatum* but differs from it in very long pedicels, 2–3 times longer than perianth (in *A. bidentatum*, they are shorter than perianth, equal or slightly longer), large number of flowers in inflorescence (not less than 20; in *A. bidentatum*, not more than 10) as well as filaments of inner stamens enlarged for 1/2–2/3 length (against 2/3–3/4 in *A. bidentatum*).

25. **A. mongolicum** Rgl. in Acta Horti Petrop. 3, 2 (1875) 160 and 10, 1 (1887) 340; Grub. Konsp. fl. MNR (1955) 93. —*A. krylovii* Sobol. in Sist. zam. Gerb. Tomsk. univ. 1–2 (1949) 9; Grub. l.c. 93. —Ic.: Rgl. l.c. 1887, tab. V, fig. 3.

Described from Mongolia. Type in Leningrad. Plate I, fig. 3; map 2.

Sandy and rocky steppes, rubble and rocky deserts, on sand, sandy knolls and rocky slopes of mountains and knolls, in saxaul scrub, sandy banks of rivers and lakes, pebble beds; from plains to lower mountain belt; sometimes ascends up to middle belt. Forms coenosis of characteristic onion desert steppes.

IA. Mongolia: *Khobd.* (Iter ad Kobdo, June 23, 1870—Kalning; basin of Uryuk-Nur lake, feather-grass desert steppe south-west of lake, July 30; Achit-Nur lake basin 10 km nor.-east of Bukhu-Muren-Gol river, rocky feather-grass desert steppe, July 31—1945, Yun.); *Mong. Alt.* (left rocky bank of Senkir river, July 17; bank of Bulugun river above Dzhirgalante river estuary, July 25; mounds along bank of Bulugun river, Aug. 4—1898, Klem.; along Saksai river, Aug. 1, 1909—Sap.; 30 km nor.-east of Bayan somon along road to Khan-Taishiri mountain range from Shargiin-Gobi, wormwood-feather-grass desert steppe on loam, Aug. 24, 1943; 35–40 km south-south-east of Tugrik-Sume, Bodkhyn-Ama area, pea shrub-feather-grass steppe, Aug. 11, 1945;

Tamchi somon, 7–8 km east of Gun-Tamga area, feather-grass desert steppe on loam, Aug. 2, 1947; Khubchin-Nuru mountains west of Adzhi-Bogdo, Tukhymyin-Khundei area, knoll slope, rubble steppe, Aug. 3, 1947; valley of Bulugun river, near winter camp in Bulugun somon, in alluvial pebble beds, Aug. 1949—Yun.; "valley of Delyun river, Sap."—Sobol. l.c.; Grub. l.c.); *Cent. Khalkha* (nor.-east of Ongiin river, on steppe, in fine rubble, July 29, 1893—Klem.; between Borokhchin lake and Sudzhi river, loamy grass steppe, July 8; peak of Delger-Khangai mountain range, July 19—1924, Pavl.; left bank of Toly, Altyn-Obo area, on loam, Aug. 5, 1925—Gus.; Tsagan-Nor lake environs, Aug. 5, 1925—Glag.; Ikhe-Tukhum-Nor environs, gully north-east of lake and its bank, June; second and third terraces of Ikhe-Tukhum-Nor lake, June—1926, Zam.; Sangin-Dalai, nor.-west. slope of south. watershed, July 14, 1926—Lis.; Khokhu-Khoshun, rocky slope, July 24, 1926—Gus.; 45–50 km south-south-west of Sorgol-Khairkhan-Ul along old road from Ulan-Bator to Dalan-Dzadagad, solonchak-like derris groves, July 16, 1943—Yun.); *East. Mong.* (Muni-Ula, July 13, 1871—Przew., syntypus!; right bank of Huang He river before Khekou town, on sand mounds, Aug. 7, 1884—Pot.; lower course of Kerulen beyond Botszanchi-Sume, 1899, Pal.; Baishintyn-Sume environs, Urgo foothill, July 17; same locality, Urgo mountain on way to Gungu-Khuduku, Aug. 18—1927, Zam.; along Yaman-Ikhe-Dulan-Khoshun road, Aug. 20; Shilin-Khoto town, on steppe, 1959—Ivan.); *Depr. Lakes* (Khirgis-Nur lake environs, Tsagan-Ulum cape on Dzabkhyn river, on sand, Aug. 9; same locality, Khara-Usu lake, Khara-Argalantu mountains, Aug. 11; nor. bank of Khara-Nor lake, Aug. 12; Un'yugut mountains on Khara-Usu lake, Aug. 14—1879, Pot.; 10 versts from Ubsa lake, near irrigation ditch, July 3, 1892—Kryl.; right bank of Bogden-Gol lake, among rocks, July 2; environs of Tsakhir-Bulak spring, July 18—1894, Klem.; sand in watershed of Tuguryuk and Dzerin rivers, Aug. 17, 1930—Bar.; 10 km nor.-nor.-west of Ulangom along road to Tsagan-Nur, desert steppe, July 29, 1945—Yun.; Dzabkhyn-Gol valley 20 km beyond Tsagan-Olom, Aug. 2, 1950—Kuznetsov); *Val. Lakes* (sand bank of Orok-Nur lake, Sept. 2, 1886—Pot.; near Orok-Nur lake, July 13, 1893, Klem.; Tuin-Gol, rocky slopes, Sept. 1; same locality, rubble-pebble bed steppe, Sept. 6—1924, Pavl.; steppe in lower courses of Baidarik river, Sept. 13, 1924—Gorbunova; Orok-Nur lake, solonetz sand, Aug. 6, 1926—Tug.; right bank of Tuin river opposite ford 15 km beyond Dzhinset somon, on sand and sand-dunes, Aug. 16, 1949—Kal.); *Gobi Alt.* (Tostu mountain range, Aug. 18; valley between Nemegetu and Tostu mountain ranges, Aug. 20—1886, Pot.; Noyan somon, sand in valley nor. of Noyan-Bogdo, July 24, 1943—Yun.; Bain-Gobi somon, between Tsagan-Gol and Ikhe-Bain-Ul, 12 km south-west of Gashun-Bulak, on slope of knoll, Aug. 1; same locality, south of Ikhe-Bain-Ula in Tavun-Khobur-Khuduk region, hillocky area on rocky slopes, Aug. 2—1948, Grub.; 40 km west of Khurmein somon along road to Bayan-Dalai, floor of extensive valley between Gurban-Saikhan and Argalantu, feather-grass-onion steppe, Aug. 15, 1962—Lavrenko); *East. Gobi* (Schabarakh-usu, on upland slopes, 1320 m, 1925—Chaney; Alkha-Khoshuin-Gobi, Aug. 13, 1927—Zam.; Argali mountain range before approaching Uda lowland, Aug. 24, 1928—L. Shastin; Delger-Khangai somon, Khoir-Ul'tszeitu area and Sharangad between Lus and Kholtu somons, Sept. 8–15, 1928—E. Kuznetsov; Bain-Dzak, Sept. 2, 1931—Krupenin; Bain-Dzak 30 km nor. of Barun-Saikhan mountain, sandy soil near saxaul forest, Sept. 25, 1931—Ik.-Gal.; 35–38 km nor.-east of Khan-Bogdo somon, sand with saxaul and wormwood, Sept. 22, 1940—Yun.; 70 km south of Sain-Shanda, lowland, Aug. 30, 1950—Ivan.; Lusain-Khudal, among chee grass scrub, July 21, 1950—Kal.; Mandal-Obo somon, along road from west into Bain-Dzak, sand ridges, July 28, 1951—Kal.); *Gobi Trans-Alt.* (Atas-Bogdo hill, east. trail of mountain and along mountain slopes, Aug. 12–13, 1943—Tsebigmid; Tsagan-Deris area, saltwort desert, July 28; Tsagan-Bogdo mountain range, upper mountain belt, rocky slopes, Aug. 4; nor. trail of Khukh-

Tumurtu mountain range near Bomiin-Khuduk well, rubble-rocky desert, Aug. 6—
1943, Yun.; Khugshu-Ula hill, nor.-west. trail, Suchzhi-Khuduk well, *Ephedra*-common
Russian thistle desert on pebble bed, Aug. 10, 1948—Grub.; 93 km nor. of An'si, broad
dry bed between knolls, July 26, 1953—Petr.); *Alash Gobi* (Kobden-Usu between
Gashiun lake and Tostu mountain range, Aug. 13, 1886—Pot.; Alashan, Yamata
gorge, south. slope, midbelt, on sandy soil, 1908—Czet.; from border of Gansu
province in Alashan up to Tsokhor-Tologoi well, Aug. 25, 1880—Przew.; Bayan-
Khoto, Tengeri sand, July 31, 1958—Petr.); *Ordos* (30 km south-east of San'shingun
town, right bank of Huang He river, sandy-pebble bed flat, July 16; 60 km west of
Ushinchi town (Ushin), overlying sand on bank of Ulibu-Nor salt lake, Aug. 2; 35 km
south-east of Khanchi-n'chi town, sandy-pebble bed flat, Aug. 7; 30 km south of Dalat
town, meadow in Huang He river valley, around Chzhandanchzhao village, Aug. 10;
24 km nor. of Dzhazak town, fixed sand, Aug. 15—1957, Petr.); *Khesi* (along Dankhe
river, 2400–2500 m, pebble bed, July 7, 1879—Pot.; Sachzhou oasis, rocky steppe,
1500–2100 m, July 24, 1895—Rob.; 38 km west of Tszyutsyuan' town in Edzin-Gol,
saltwort desert, Oct. 7, 1957—Yun. et al.; 10 km west of Tszyutsyuan' town, pebble
bed-loamy sand foothill flat, July 25, 1958—Petr.).

IIA. **Junggar**: *Cis-Alt*. (Fuyun' [Koktogai]—Ukagou, No. 1679, Aug. 11, 1956—
Ching); *Jung. Gobi* (nor.: Mukurtai area west of Ulyungur lake, on sand, Aug. 11;
south-east. bank of Ulyungur lake, on sand, Aug. 15—1876, Pot.; right bank of
Urungu river 6 km nor. of Ertai along road to Koktogai, desert along the lower part of
mountain trails, July 14, 1959—Yun.; cen.: 60 km south of Ertai, hillocky desert area,
No. 1288, July 16, 1959—Yun. and I.-F. Yuan'; south.: Tsitai [Guchen], on sand, No.
2347, Sept. 26; from Gan'khetsz to Tsitai, No. 5132, Sept. 29—1957, Kuan; 41 km east
of Muleikhe town, No. 2090, Sept. 23; Tsitai district, from Meido to Beidashan' in
south, in desert, No. 165, Sept. 26—1957, Kuan; east.: cliffs around Kyup spring, Aug.
8, 1898—Klem.); *Zaisan* (between Ch. Irtysh and Saur, rubble steppe, Aug. 21, 1906—
Sap.).

IIIA. **Qinghai**: *Nanshan* (Kuku-Usu river, Aug. 4, 1879—Przew.; 5 km south of
Aksai, high foothills of Altyntag mountain range, Aug. 2, 1958—Petr.).

General distribution: East. Sib. (Tuva), Nor. Mong. (Khangai—south).

Note. *A. krylovii* Sobol. l.c. was described from East. Siberia (valley of Ulu-Kkhem
river, near Eleges estuary in Tuva Autonomous Soviet Socialist Republic) and recorded
by Sobolevskaja and Grubov (l.c.) for the Mongolian People's Republic [Khobd., Mong.
Alt. (Delyun), Great Lakes Depression (Ubsa-Nur) areas]. A study of an authentic
specimen of *A. krylovii* (vicinity of Chadan experimental station, Tuva Autonomous
Province, chee grass scrub, June 24–28, 1947, Samoilova and Skvortzova) supplied by
P.N. Krylov Herbarium (Tomsk University) showed this species to be totally similar to
A. mongolicum Rgl. K.A. Sobolevskaja differentiates *A. krylovii* from *A. mongolicum* in
filaments of stamens as long as perianth lobes or 1.5–2 times shorter, smaller size of
flowers (5–7 (8) mm long) as well as in very low scapes. However, the length ratio of
filaments to perianth indicated by Sobolevskaja reflects the variation range of this
characteristic in *A. mongolicum* throughout its distribution range. According to Regel's
diagnosis, filaments of *A. mongolicum* are slightly longer than 1/2 perianth lobes. A
study of the numerous specimens of *A. mongolicum* showed that filaments in this
species may even be shorter than perianth lobes, sometimes almost as long, while
perianth size varies from (5) 6 to 9 mm but more often 6–8 mm long. Perianth lobes
vary in shape from elliptical to broadly and oblong-elliptical.

26. **A. prostratum** Trev. in Ind. Sem. Hort. Wratisl. (1822) 16; Vved. in
Fl. SSSR, 4 (1935) 166; Kitag. Lin in Fl. Mansh. (1939) 131; Grub. Konsp, fl.

MNR (1955) 94. —*A. stellerianum* γ. *prostratum* (Trev.) Rgl. in Acta Horti Petrop. 3, 2 (1875) 150 and 10, 1 (1887) 337.

Described from East. Siberia. Type in ? Bratislava. Plate I, fig. 4.

Steppe rubble and rocky slopes, pebble beds; foothills and lower mountain belt, rarely middle and upper belts.

IA. **Mongolia:** *Khobd.* (mountains along left bank of Kharkhira river, moist sites, July 11, 1879—Pot.; 3–4 km west of Ulan-Daba pass along road to Tsagan-Nur from Ulangom, dry montane steppe, July 29, 1945—Yun.); *Mong. Alt.* (south of Tsitsiriin-Gol, in alpine belt, on pebbles, July 9; Taishiri-Ol, dry mountain slope, July 13—1877, Pot.; 10 km south-east of Yusun-Bulak, central part of trail of Khan-Taishiri mountain range, forb-wheat grass-feather-grass steppe, July 14; Tamchi somon, 3–4 km south-west of Tamchi-Nur lake, slopes of mud cones, July 17; same locality, Bus-Khairkhan mountain range, upper part of mountain trail, forb-wormwood steppe, July 17—1947, Yun.); *Cen. Khalkha* (vicinity of Orgochen-Sume monastery, granite outcrops, Aug. 23, 1925—Krasch. and Zam.; south-east. Khangai foothills, Khai area, July 26, 1926—Gus.; 30 km west of Tsagan-Oho somon, forb-feather-grass-sheep's fescue associations on tops of knolls, Aug. 8—1956, Dashnyam); *East. Mong.* (between Dolon and Dzhirgalante, July 29, 1898—Zab.); *Gobi Alt.* (Dundu-Saikhan mountains, July 9, 1909—Czet.; nor. offshoots of Ikhe-Bogdo mountain range, July 10, 1926—Kozlova; pass between Dzun-Saikhan and Dundu-Saikhan mountains, grass steppes along floor of dry gully in lower mountain belt, July 22, 1943; same locality, rocky slopes, July 22, 1943; west. part of Dzun-Saikhan mountain range, steppe slope in upper belt, June 19, 1945—Yun.: 35 km from Dalan-Dzadagad, rocky steppe on south. slope of Dundu-Saikhan mountain range, July 20, 1950—Kal.); *East. Gobi* (between Chzhalat and Khachzhabtsa, July 4, 1898—Zab.; on way from Alashan to Urgu [Ulan-Bator], along road from Mandalyn-Gobi to Khara-Tologoi area, June 8, 1909—Czet.).

General distribution: East. Sib. (south.), Nor. Mong.

27. **A. rubens** Schrad. ex Willd. Enum. pl. Hort. Berol. 1 (1809) 360; Vved. in Fl. SSSR, 4 (1935) 166; Fl. Kazakhst. 2 (1958) 153; Vved. in Opred. rast. Sr. Azii [Key to Plants of Mid. Asia], 2 (1971) 61. —*A. stellerianum* auct. non Willd.: Rgl. in Acta Horti Petrop. 3, 2 (1875) 145, p.p., quoad syn. *A. rubens* Schrad; l.c. 10, 1 (1887) 337, p.p., quoad var. α. and β.; Kryl. Fl. Zap. Sib. 3 (1929) 620, p.p., quoad syn. *A. rubens* Schrad. —Ic.: Fl. Kazakhst. 2, Plate 13, fig. 6.

Described from Siberia. Type in Berlin (B) ?

Rocky slopes; lower and middle mountain belts.

IIA. **Junggar:** *Jung. Alatau* (in vicinity of Toli, on slope, No. 2147, Aug. 8, 1957—Kuan).

General distribution: Jung.-Tarb.; West. Sib. (south.), East. Sib. (? south.).

28. **A. senescens** L. Sp. pl. (1753) 299; Rgl. in Acta Horti Petrop. 3, 2 (1875) 137 and 10, 1 (1887) 336; Franch. Pl. David. 1 (1884) 304; Palib. in Tr. Troitskosavsk.-Kyakht. otdel. RGO [Russian Geographic Society], 7, 3 (1904) 42; Forbes and Hemsley, Index Fl. Sin. 3 (1905) 125; Kryl. Fl. Zap. Sib. 3 (1929) 617; ? Pampanini, Fl. Carac. (1930) 88; Vved. in Fl. SSSR, 4 (1935) 169; Kitag. Lin. Fl. Mansh. (1939) 132; Grub. Konsp. fl. MNR (1955) 94; Fl. Kazakhst. 2 (1958) 154; Vved. in Opred. rast. Sr. Azii [Key to Plants

of Mid. Asia], 2 (1971) 61. —*A. senescens* var. *fallax* Trautv. in Acta Horti Petrop. 1, 2 (1872) 193.

Described from Europe and Siberia. Type in London (Linn.). Plate I, fig. 1.

Steppes, steppe slopes, meadows, river and lake valleys. From plains to lower mountain belts, rarely middle belt.

IA. **Mongolia:** *Khobd.* (Iter ad Kobdo, June 20, 1870—Kalning); *Mong. Alt.* (Buyantu river, in floodplain, Aug. 27, 1930—Bar.; Bulugun somon, lower course of Turgen-Gol river, dry meadow, July 7, 1947—Yun.); *Cen. Khalkha* (Budzynkhe sand, Aug. 11, 1925—Gus.; Ikhe-Tukhum-Nor lake environs, Ulanche-Delger mountain, June; same locality, Modkho mountain, June; same locality, Ongon-Khairkhan hill, June—1926, Zam.; 10 km west of Tumen-Delger somon, Morzan-Khoolai, feather-grass-forb-wild rye steppe, Aug. 8, 1956—Dashnyam; 23 km south of Munku-Khan somon, on ridge in forb-grass associations, July 29, 1962—Yun.); *East. Mong.* (Ourato centr., July 1866—David.; steppe south of Bain-Nor lake, June 20; Kulun-Buir-Norsk plain, steppe between Abder river and Borol'-Dzhitu area, June 26; same locality, Lukh-Sume monastery, July 3, 1899—Pot. and Sold.; 44 km west-north-west of Cheibalsan, feather-grass-wild rye steppe, Aug. 1; Khuntu somon, 28–30 km south-east of Bain-Buridu, forb-grass steppe, Aug. 5; same locality, 30 km east of Bain-Tsagan, feather-grass-forb steppe, Aug. 6; same locality, 35 km east of Bain-Tsagan, meadow steppe, Aug. 6; same locality, 18 km south-east of Bain-Tsagan, wild rye-alkali grass meadow on gully floor, Aug. 6; 55 km east of Erentsab, Shavorte area, rocky slopes, Aug. 19; Khalkha-Gol somon, 30 km from Bain-Buridu, forb meadow, Aug. 18—1949, Yun.; Khailar town, sand-dunes, No. 2875, 1954—Wang; 30 km south-east of Choibalsan, hillocky flat, Aug. 24; Choibalsan somon, 47 km east of Chingiskhan shaft, forb steppe [undated]—1954, Dashnyam; 13 km from Kherulen-Bayan somon, in tansy-feather-grass association on slope, Aug. 10; 13 km south of Kherulen-Bayan somon, top of Gomyn-Shile ridge, forb-snakewort-feather-grass association, Aug. 10—1956, Dashnyam; nor. of Choibalsan somon, south. foothill of Delger-Moun hill, Aug. 29, 1959—Dashnyam; Khailar town vicinity, in steppe, 1959—Ivan.; Choibalsan somon, 12 km nor.-east of Enger-Shand toward Khabirga mountains, July 22, 1962—Lavrenko; 25 km nor. of Khamar-Daba, feather-grass-koeleria-wormwood-snakewort association, June 26, 1971—Dashnyam, Rachk. et al.; "Inter Kulussutajewsk et Dolon-Nor, 1870, Lom."—Trautv. l.c.; "Between Dolon and Dzhirgalante, between Chzhalate and Khachzhabtsa, 1898, Zab."—Pal. l.c.); *Depr. Lakes* (Barun-Turun valley near emergence of Kharkhira river, June 3, 1931—Bar.; "Tannu-Ol"—Grub. l.c.); *Gobi Alt.* (Ikhe-Bogdo mountain range, Bityuten-Ama creek valley, mountain slopes, Aug. 12, 1927—Simukova; Dundu-Saikhan mountains, Aug. 17, 1931—Ik.-Gal.; pass between Dundu-Saikhan and Dzun-Saikhan mountains, floor of arid gully, Aug. 22, 1943—Yun.); *Ordos* (Baga-Edzhin-Khoro, Aug. 18; Ulan-Morin river, Aug. 23—1884, Pot.).

IIA. **Junggar:** *Jung. Gobi* (nor.: Altay-Burchum, No. 3214, Sept. 16, 1956—Ching).

General distribution: Jung.-Tarb. (Tarbagatai and Saur); Europe, West. Sib. (south., including Altay), East. Sib. (south.), Far East, Nor. Mong., China (Dunbei, North).

Note. Polymorphous species that served as the basis for describing some varieties. Specimens cited above are quite stable in characteristics. Variation is mainly seen in leaf breadth and shape of perianth lobes—from oblong and oblong-ovate (most common) to ovate or broadly ovate (rare). Perianth lobes of latter 2 types are seen in plants of Khobdos region, Mongolian (Buyantu river) and Gobi Altay, It is possible that specimens of *A. senescens* with broad perianth lobes represent transitional forms with species of very closely related *A. prostratum* Trev. (see above). The specimen of *A. senescens* from Junggar Gobi (northernmost) bears teeth on filaments of inner stamens, a characteristic rarely found in the present species.

29. **A. stellerianum** Willd. Sp. pl. 2, 1 (1799) 82; Rgl. in Acta Horti Petrop. 3, 2 (1875) 149, p.p., excl. syn. *A. rubens* Schrad. and *γ. prostratum* (Trev.) Rgl.; Franch. Pl. David. 1 (1884) 304; Sap. Mong. Alt. (1911) 388; Kryl. Fl. Zap. Sib. 3 (1929) 620, p.p., excl. syn. *A. rubens*; Vved. in Fl. SSSR, 4 (1935) 165; Grub. Konsp. fl. MNR (1955) 94. —*A. senescens* var. *flavescens* Rgl. in Acta Horti Petrop. 10, 1 (1887) 336.

Described from Siberia. Type in Berlin (B) ?

Arid rocky and rubble steppe slopes of mountains.

IA. Mongolia: *Khobd.* (Iter ad Chobdo, Altyntschetsche, July 9, 1870—Kalning; Kharkhira river valley near Turgun' river estuary, on moist sand, July 21, 1879—Pot.; Turgen' mountain range, Turgen'-Gol gorge 2 km from estuary, right bank terrace, forb-sheep's fescue steppe, July 16, 1971—Grub. et al.; "Kalgutty"—Sap., l.c.); *Depr. Lakes* (Ubsa-Nur lake environs [without exact date]—1879, Pot.; "South. slope of Tannu-Ol"—Grub. l.c.); *East. Mong.* ("Chaîne de l'Ourato a'Mao Mingan, July 1866, David"—Franch. l.c.).

General distribution: West. Sib. (south., including Altay), East. Sib. (south.).

30. **A. tenuissimum** L. Sp. pl. (1753) 301; Rgl. in Acta Horti Petrop. 3, 2 (1875) 157, p.p., excl. var. *anisopodum* (Ledeb.) Rgl. and *A. bidentatum* Fisch.; ibid 10, 1 (1887) 341, quoad *α. typicum*, p.p.; Franch. Pl. David. 1 (1884) 305; Forbes and Hemsley, Index Fl. Sin. 3 (1905) 125; Sap. Mong. Alt. (1911) 388; Danguy in Bull. Mus. nat. hist. natur. 20 (1914) 140; Kryl. Fl. Zap. Sib. 3 (1929) 623, p.p.; Vved. in Fl. SSSR, 4 (1935) 173, p.p.; Kitag. Lin. Fl. Mansh. (1939) 132; Walker in Contribs, U.S. Nat. Herb. 28 (1941) 602; Dashnyam in Bot. zh. 50, 11 (1965) 1640; Grub. in Novo. sist. vyssh. rast. 9 (1972) 276.

Described from Siberia. Type in London (Linn.). Plate I, fig. 5.

Arid and semidesert steppes along slopes of mountains and mud cones, rocky and rubble slopes; up to middle mountain belt.

IA. Mongolia: *Mong. Alt.* (Tsetseg-Nur somon, Tsetseg-Nur basin, desert steppe on gentle slope, Aug. 11; Tonkhil somon, east. slope toward Tonkhil-Nur basin, montane feather-grass desert steppe, Aug. 13—1945, Yun.; Tamchi somon, 5 km north-west of Tamchi-Daba, wormwood-forb montane steppe, July 16; Tonkhil somon 6 km south of Tonkhil-Nur lake along road to Tamchi somon, desert steppe on inclined trough, July 16; same locality, pass through Dobtsig-Khuren-Nur, steppe along hillocky area, Aug. 11—1947, Yun.; "Tsagan-Gol river; Dain-Gol lake; Sumdairyk river"—Sap. l.c.); *Cent. Khalkha* ("Bulgan somon in Ubur-Urtu region, on south. dry slopes of knolls in pea shrub-wild rye steppe"—Dashnyam, l.c., Grub. l.c.); *East. Mong., Gobi Alt.* (steppe on south. slope of Dundu-Saikhan mountain range, along road from pass to Khongor somon, July 8; Gurban-Saikhan mountain range, pass between Dzun-Saikhan and Dundu-Saikhan mountains, peak east of Dundu-Saikhan mountain border, on rocks, July 22—1943, Yun.; Dzun-Saikhan mountain range, middle belt, upper boundary of juniper groves, June 19, 1945—Yun.).

General distribution: West. Sib. (south. including Altay), East. Sib. (south.), ? Far East, China (Dunbei).

Note. A study of the photograph of type specimen of *A. tenuissimum* showed an extremely sparse inflorescence consisting of 3 flowers with equal pedicels. The original diagnosis of this species contains a reference to the work of J.G. Gmelin [J.G. Gmel., Fl. Sib. I (1747) 61, tab. 15, figs. 2, 3]. Gmelin's figure depicted 2 plants: low-growing, with nearly equal pedicels and a much taller one with unequal pedicels. In both plants, the inflorescence contains few flowers. Plants with unequal pedicels, as in onion depicted in Gmelin's figure, were described by K.F. Ledebour [Ledeb., Fl. Ross. 4, 1 (1852) 183] as *A. anisopodum*. Apart from pedicels of varying length, the original author characterises this species with many-flowered inflorescence (20–40 flowers) and oblong perianth lobes (an additional fairly constant characteristic is the presence on perianth lobes of cuneate thickening of denser consistence than elsewhere on the surface). Ledebour placed in *A. tenuissimum*, plants with few-flowered (4–10 flowers) inflorescence and rounded-elliptical outer and obovate inner perianth lobes. A study of the vast herbarium material showed, in many cases, a correlation between number of flowers in inflorescence, nature of pedicels and shape of perianth lobes. On the basis of the sum total of these characters, some plants could be placed in *A. anisopodum* and others (very few) in *A. tenuissimum* (But only in sense of Ledebour, because, it cannot be confidently stated that *A. tenuissimum* L.-*A. tenuissimum* sensu Ledeb. without a study of the detailed structure of the flowers). The number of flowers in the inflorescence varies among *A. anisopodum*: quite often plants with a small number of flowers and nearly equal pedicels are found. There are also intermediates between *A. tenuissimum* and *A. anisopodum*.

31. **A. tibeticum** Rendle in J. Bot. (London) 44 (1906) 41.
Described from Tibet. Type in London (K).
Alpine belt.

IIIB. Tibet: *South.* ("Karo La, fifteen miles from Lhasa, Aug. 13, 1878, Dungboo"—Rendle, l.c.; "Near Maku La, July-Aug. 1903, Younghusband"—Rendle, l.c.; "Karo La Pass, about 4950 m, July 1904, Walton"—Rendle, l.c.).
General distribution: Himalayas.

Note. Rendle relates *A. tibeticum* to the Himalayan species *A. sikkimense* Baker, distinguishing it from the latter in very small sky-blue (not lilac-purple), sub-globose-campanulate (not campanulate) flowers and spatulately enlarged filaments of inner stamens.

Section **Oreiprason** F. Herm.

32. **A. blandum** Wall. Pl. Asiat. rar. 3 (1832) 38; Hemsley in J. Linn. Soc. (London) Bot. 30 (1894) 123, 124; Strachey, Catal. (1906) 188; Pampanini, Fl. Carac. (1930) 87; Vved. in Opred. rast. Sr. Azii [Key to Plants of Mid. Asia], 2 (1971) 63. —*A. polyphyllum* auct. non Kar. et Kir.: Vved. in Fl. SSSR, 4 (1935) 176, p. min. p., quoad pl. pamir. nonnull.; idem in Fl. Tadzh. 2 (1963) 310, quoad pl. pamir. nonnull.; Ikonn. Opred. rast. Pamira [Key to Plants of Pamir] (1963) 88, p.p. —Ic.: Wall., l.c. tab. 260.
Described from Himalayas. Type in London (BM).

IIIB. Tibet: *Chang Tang* ("Kuen-lun Plains at about 5100 m, Picot"—Hemsley, l.c.; "Lingzi-tang ed Akhsáe-cin, ca. 5175 m, Picot"—Pampanini, l.c.; "Valle Zug Shaksgam, 4325 m, Clifford"—Pampanini, l.c.).

General distribution: East. Pam.; Mid. Asia (West. Pamir), Himalayas (west., east., Kashmir).

Note. The species is very closely related to *A. carolinianum* DC. (see below). A.I. Vvedensky (l.c. 1971) differentiates *A. blandum* from *A. carolinianum* (= *A. polyphyllum* Kar. et Kir.) in characteristics given here in the key. Wendelbo [in Reching. f. Fl. Iran. 76, 1, 4 (1971) 14] treated *A. blandum* as a synonym for *A. carolinianum*.

33. **A. carolinianum** DC. in Redouté, Liliac. 2 (1804) tab. 101; Wendelbo in Reching. f. Fl. Iran 76, 1, 4 (1971) 14. —*A. polyphyllum* Kar. et Kir. in Bull. Soc. natur. Mosc. 15, 3 (1842) 509; Rgl. in Acta Horti Petrop. 3, 2 1875) 129 and 10, 1 (1887) 328; Danguy in Bull. Mus. nat. hist. natur. 20 (1914) 140; Paulsen in Hedin, S. Tibet, 6, 3 (1922) 89; Pampanini, Fl. Carac. (1930) 88; Hand.-Mazz. in Österr. Bot. Z. 79 (1930) 39; Vved. in Fl. SSSR, 4 (1935) 176, p. max. p., excl. pl. pamir. nonnull.; Persson in Bot. Notiser, 4 (1938) 277; Fl. Kirgiz. (1951) 66; Fl. Kazakhst. 2 (1958) 159; Fl. Tadzh. 2 (1963) 310, p.p. excl. pl. pamir. nonnull.; Ikonnik. Opred. rast. Pamira [Key to Plants of Pamir] (1963) 88, p.p.; Vved. in Opred. rast. Sr. Azii [Key to Plants of Mid. Asia], 2 (1971) 63. —*A. thomsonii* Baker in J. Bot. (London) 12 (1874) 294. —**Ic.:** Redouté, l.c. tab. 101; Fl. Tadzh. 2, Plate 57, fig. 1 (sub nom. *A. polyphyllum*).

Described from specimens grown from bulbs of unknown origin. Type in Paris (P).

Rocky and rubble slopes in subalpine and alpine belts.

IB. Kashgar: *West.* (Kuen'-Lun', upper Tiznaf river, 25 km beyond Kyude settlement along Tibet highway to Saryk-Daban, 3900 m, No. 242, June 1, 1959—Yun. et al.); *South.* ("Nordfluss des Kisil-dawan, felsiges Tal des Kurab Su, 2950 m, June 18, 1906, Zugmayer"—Hand.-mazz. l.c.).

IIA. Junggar: *Tien Shan* (Sairam lake, Talki brook, July 19; south-east. bank of Sairam lake, July 23—1877, A. Reg.; Kash river, Aug. 18, 1878—A. Reg.; Davanchin intermontane valley, Ulumbai area, 20–25 km south of Urumchi, along solonchak meadow on bank of spring, June 2; same locality, Manas river basin, Ulan-Usu river valley at confluence with Dzhirtas river, subalpine belt, southern rocky slope, July 19— 1957, Yun. et al.; between Ulastai and Yakou, among stones, No. 3984, Aug. 30, 1957—Kuan; Kalan'gou village in Turfan region, 2500 m, No. 5782, June 23; 10 km north of Kalan'gou village, 2600 m, No. 5801, June 23—1958, Lee and Chu); "Sairam-Nor, July 18, 1895, Chaff."—Danguy, l.c.); *Jung. Gobi* (Beidashan', among rocks, 3300 m, No. 2928, Sept. 29, 1957—Lee et al.).

IIIC. Pamir (Ulug-Rabat pass, along sandy-clayey descent, July 23, 1909—Divn.; Yazlek river valley, right tributary of Shinda river, Aug.-Sept. 1941; Pil'nen river, 3000– 4000 m, June 30, 1942; source of Kashka-Su river, 4200–5500 m, July 5, 1942—Serp.; Sarykol mountain range, 10–12 km south of Ulug-Rabat pass along road to Tashkurgan from Kashgar, small block of prickly thrift and crazy weed, 4100 m, June 14, 1959—Yun.; "Eastern Pamir, sandhills at the eastern shore of Little Kara-kul, 3720 m, July 17, 1894, Hedin"—Paulsen, l.c.; "Jerzil, growing in big tussocks on hillsides, about 3500 m, July 19, 1930"—Persson, l.c.).

General distribution: Jung.-Tarb., Nor. and Cen. Tien Shan, East. Pam.; Fore Asia (Afghanistan, Pakistan), Mid. Asia (West. Tien Shan, Pam.-Alay), Himalayas.

34. A. chrysanthum Rgl. in Acta Horti Petrop. 3, 2 (1875) 91 and 10, 1 (1887) 313; Walker in Contribs. U.S. Nat. Herb. 28 (1941) 602.

Described from Qinghai (Nanshan). Type in Leningrad.

Meadows in alpine belt.

IIIA. Qinghai: *Nanshan* (South Tetung mountain range, alpine meadows, July 25, 1872—Przew., typus!; nor. slope of South Tetung mountain range, alpine meadows, 3000–3600 m, July 31, 1880—Przew.; North Tetung mountain range, Cherik pass, Aug. 8, 1890—Gr.-Grzh.; Ganshig river sources, tributary of Peishikhe river, moraine, 3900–4300 m, Aug. 18, 1958—Dolgushin; "Tai Hua; Ta Pan Shan La Chi Tzu Shan; scattered on exposed moist steppes or densely bushy mountain-sides, 1923, Ching"—Walker, l.c.).

General distribution: China (North-West, Cent., South-West).

35. A. chrysocephalum Rgl. in Acta Horti Petrop. 10, 1 (1887) 335; Forbes and Hemsley, Index Fl. Sin. 3 (1905) 121; Hao in Bot. Jahrb. 68 (1938) 587. —Ic.: Rgl. l.c. tab. 3, fig. 1.

Described from Qinghai. Type in Leningrad.

Meadows; alpine belt.

IIIA. Qinghai: *Nanshan* (mountain range between Nanshan and Donkyr, along Rako-Gol river, alpine meadow, 3000–3300 m, Aug. 22, 1880—Przew., typus!).

IIIB. Tibet: *Weitzan* (Yantszytszyan river basin, Chamudug-La pass, 4710 m, July 26, 1900—Lad.; "Den Selgen-Gebirge, 4800 m, auf Kies, Aug. 21, 1930; "Amne-Matchin, 4350 m, Sept. 1, 1930, südlich von Veridagan-zo am Ulter, Sept. 2, 1930"—Hao, l.c.).

General distribution: endemic.

36. A. condensatum Turcz. in Bull. Soc. natur. Mosc. 27, 2 (1885) 121; Rgl. in Acta Horti Petrop. 3, 2 (1875) 105; Franch. Pl. David. 1 (1884) 304; Palib. in Tr. Troitskosavsko-Kyakht. otdel. RGO, 7, 3 (1904) 41; Forbes and Hemsley, Index Fl. Sin. 3 (1905) 121; Vved. in Fl. SSSR, 4 (1935) 182; Kitag. Lin. Fl. Mansh. (1939) 131; Grub. Konsp. fl. MNR (1955) 92.

Described from East. Siberia. Type in Leningrad.

Steppes.

IA. Mongolia: *East. Mong.* (Borogolo, June 31; Kalgan gorge, Aug. 23, 1898—Zab.; Kulun-Buir-Nur plain, Abder river, June 24 and 25; same locality, Bilyutai, July 8; same locality, Ludzhin [Lodzin]—Kharaul, clayey-sandy arid soil, July 9—1899, Pot. and Sold.; Yugodzyr, Sept. 30, 1928—L. Shastin; Khalkhin-Gol somon, 10 km west of Bain-Buridu, wheat grass-wormwood steppe, Aug. 5; same locality, 28–30 km south-east of Bain-Buridu, forb-grass steppe, Aug. 5; Khuntu somon, 17–20 km east-south-east of Bain-Tsagan, feather-grass-tansy steppe on loam, Aug. 6—1949, Yun.; Shilin-Khoto, steppe, 1957—Ivan.; Khalkhin-Gol somon, forb-wormwood association, July 24, 1962—Dashnyam; "Boro-Gol, 1898, Zab."—Palib. l.c.).

General distribution: East. Sib. (Dauria), Far East, Nor. Mong. (Fore Hing.), China (Dunbei, North), Korean peninsula.

37. **A. consanguineum** Kunth. Enum. pl. 4 (1843) 431; Rgl. in Acta Horti Petrop. 3, 2 (1875) 131. —*A. consanguineum* var. *roseum* Rendle in Moore and Rendle in J. Bot. (London), 38 (1900) 429; Deasy, In Tibet and Chin. Turk. (1901) 404.

Described from Himalayas. Type in Berlin (B) ?

Alpine belt.

IIIB. Tibet: *Chang Tang* ("Sarok Tuz valley, 3900 m, 1898, Deasy"—Rendle, l.c.; Deasy, l.c.).

General distribution: Himalayas.

Note. The cited specimen belongs to *A. consanguineum* var. *roseum* Rendle, l.c., described from Tibet with type in London (BM). According to the original author, this variety differs from the type one (not yellow) in pink coloration of the perianth. As far as could be judged from the original diagnosis, *A. consanguineum* is close to *A. chrysanthum* Rgl., the former differing from the latter in much smaller but equal perianth lobes, fibrous tunic of bulbs and fasciculate inflorescence.

38. **A. globosum** M.B. ex Redouté, Liliac. 3 (1807) tab. 179; Rgl. in Acta Horti Petrop. 3, 2 (1875) 197, quoad *α. typicum*, and 10, 1 (1887) 351, p.p.; Vved. in Fl. SSSR, 4 (1935) 185; Fl. Kazakhst. 2 (1958) 165; Vved. in Opred rast. Sr. Azii [Key to Plants of Mid. Asia], 2 (1971) 66. —*A. globosum* f. *dilute-roseum* Kryl. Fl. Zap. Sib. 3 (1929) 632. —*A. dshungaricum* Vved. in Opred. rast. Sr. Azii [Key to Plants of Mid. Asia], 2 (1971) 66, nom. invalid.

Described from Nor. Caucasus. Type in Paris (P).

Rocky hill slopes; lower and middle mountain belts.

IIA. Junggar: *Cis-Alt.* (Kandagatai river, 1876—Pot.); *Jung. Alatau* (Mai-Kapchagai mountain, rocky slopes, June 6, 1914—Schischk.; Maili mountain range, upper Laba river, desert mounds, July 6, 1905—Sap.; Dzhair mountain range, 1–2 km nor. of Otu settlement along highway to Toli and Chuguchak through Dzhair pass, steppe belt, along granite rocks, Aug. 4; Shuvutin-Daba pass nor. of Sairam-Nur basin, meadow steppe along ridge of mountain range, Aug. 18—1957, Yun. et al.).

General distribution: Fore Balkh., Jung.-Tarb., Europe (south. European part of USSR), Caucasus (Nor.), West. Sib. (south.).

Note. In my opinion the specimens cited above (barring A.A. Yunatov's collection), are indistinguishable from those of *A. globosum* from other parts of its distribution range. A.I. Vvedensky (l.c. 1971) cites *A. globosum* only for Betpakdala (Bektau-Ata). For Zaisan basin, Tarbagatai and Junggar Alatau, he cites another species for the first time—*A. dshungaricum* Vved. [l.c. (1971) 66]. In all probability, the latter is based on *A. globosum* f. *dilute-roseum* Kryl. (l.c.). A.I. Vvedensky distinguishes (in the key) *A. dshungaricum* from *A. globosum* in pink coloration of the perianth; according to Vvedensky, it is purple in *A. globosum*. As far as could be judged from herbarium material, however, the perianth is usually pink in *A. globosum* plants collected in Caucasus and southern European part of USSR. The species rank of *A. dshungaricum* can hardly be acknowledged. It must be stated that the name *A. dshungaricum* is unsupported as nomenclatural type has not been cited in the publication.

Specimens collected by A.A. Yunatov differ from type specimens of *A. globosum* in elongated, gradually acuminate perianth lobes and perhaps represent *A. goloskokovii* Vved. [l.c. (1971) 313, 65], endemic in Junggar Alatau.

39. **A. hymenorrhizum** Ledeb. Fl. alt. 2 (1830) 12; Rgl. in Acta Horti Petrop. 3, 2 (1875) 131, p.p., excl. *β. tianschanicum* (Rupr.) Rgl. 10, 1 (1887) 332; Danguy in Bull. Mus. nat. hist. natur. 20, 3 (1914) 139; Kryl. Fl. Zap. Sib. 3 (1929) 615, p.p., excl. syn. *A. kaschianum* Rgl.; Vved. in Fl. SSSR, 4 (1935) 176; Fl. Kirgiz. 3 (1951) 66; Grub. Konsp. Fl. MNR (1955) 93; Fl. Kazakhst. 2 (1958) 160; Fl. Tadzh. 2 (1963) 312; Ikonn. Opred. rast. Pamira [Key to Plants of Pamir] (1963) 88; Vved. in Opred. rast. Sr. Azii [Key to Plants of Mid. Asia], 2 (1971) 63. —Ic.: Fl. Kirgiz., 3, Plate 13.

Described from Altay. Type in Leningrad.

Melkozem and rocky slopes; from lower to upper hill belts.

IA. **Mongolia:** *Mong. Alt.* ("Altai, environs de Kobdo, Aug. 21, 1895, Chaff."— Danguy, l.c.).

IIA. **Junggar:** *Tarb.* (south. trail of Urkashar mountain range, east. fringe of sasa strip along road to Khobuk from Chuguchak, along short-grass meadows, Aug. 5, 1957—Yun. et al.); *Tien Shan* (Aktash in Dzhagastai mountains, Aug. 11, 1877—A. Reg.; mountains around Nilka, 1500 m, June; Karagol river near pass into around Nilka, 2700 m, June 6—1879, A. Reg.; Ukagou south of Fuyuan', 1100 m, No. 1758, Aug. 11, 1956—Ching; Syata, near water, No. 4697, Aug. 27, 1957—Kuan; "Sairam-Nor, montagnes, July 23, 1895, Chaff."—Danguy, l.c.).

General distribution: Jung.-Tarb., Nor. and Cent. Tien Shan, East. Pam.; Mid. Asia (West. Tien Shan, Pam.-Alay), West. Sib. (Alay).

40. **A. kaschianum** Rgl. in Acta Horti Petrop. 10, 1 (1887) 338; Vved. in Fl. SSSR, 4 (1935) 177; Fl. Kirgiz. 3 (1951) 66; Fl. Kazakhst. 2 (1958) 160; Vved. in Opred. rast. Sr. Azii [Key to Plants of Mid. Asia], 2 (1971) 63. —Ic.: Rgl. l.c. tab. 3, fig. 2.

Described from Junggar. Type in Leningrad.

Melkozem slopes in river valleys; middle and upper mountain belts.

IIA. **Junggar:** *Tien Shan* (valley of Sumbe river, 1800–2100 m, July 29, 1877—Fet.; left bank of Kash river, 2700 m, July 15, 1879—A. Reg.; lectotypus!, Arystan-Daban on Kash river, Aug. 7, 1879—A. Reg.; east of Syata, 2050 m, No. 1317, Aug. 11, 1957—Kuan; valley of M. Yuldus, on alluvium, 2550 m, No. 6371, Aug. 2, 1958—Lee and Chu (A.R. Lee (1959)).

General distribution: Cent. Tien Shan.

Note. This species, very closely related to *A. hymenorrhizum* Ledeb., requires further investigation.

41. **A. korolkowii** Rgl. in Acta Horti Petrop. 3, 2 (1875) 158; Regel in Izv. Obshch. lyubit. estestvozn. antrop. etnogr. 21, 2 (1876) 79; Vved. in Fl. SSSR, 4 (1935) 187; Fl. Kirgiz. 3 (1951) 77; Fl. Kazakhst. 2 (1958) 166; Vved. in Opred. rast. Sr. Azii [Key to Plants of Mid. Asia], 2 (1971) 67. —*A. oliganthum* var. *elongatum* Kar. et Kir. in Bull. Soc. natur. Moscou 15,

3 (1842) 511. —*A. moschatum* β. *dubium* and γ. *brevipedunculatum* Rgl. in Acta Horti Petrop. 6, 2 (1880) 522, 523. —*A. moschatum* auct. non L.: Rgl. in Acta Horti Petrop. 10, 1 (1887) 325, p.p., quoad syn. *A. korolkowii.* —Ic.: Regel, l.c. 1876, tab. 12, figs. 13–15; Fl. Kirgiz. 3, Plate 9, fig. 3; Fl. Kazakhst. 2, Plate 12, fig. 8.

Described from Kirghizstan (Cent. Tien Shan). Type in Leningrad.

Arid steppe slopes; lower mountain belt.

IIA. Junggar: Tien Shan (Maralty, toward Muzart gorge, 1800 m, Aug. 1 [1877]—Fet.; Bain-Gol river in upper courses of Tekes river, July 27, 1893—Rob.; between Ili and Dzhagastai, on slope, No. 3149, Aug. 7; 8 km east of Chzhaos [Kalmak-Kure], intermontane depression, No. 3236, Aug. 11; mountains south of Sin'yuan' [Taldy] town, on slope, No. 3718, Aug. 22—1957, Kuan; Ketmen' mountain range, 1 km before Sarbushin settlement along road to Ili from Kzyl-Kure, steppe belt, on south. rocky slope, No. 1558, Aug. 21, 1957—Yun. et al.).

General distribution: Jung.-Tarb. (Jung. Alatau), Cent. Tien Shan; Mid. Asia (Alay valley).

42. **A. megalobulbon** Rgl. in Acta Horti Petrop. 6 (1879) 526 and 10, 1 (1887) 333.

Described from Junggar (Tien Shan). Type in Leningrad.

Middle mountain belt.

IIA. Junggar: *Tien Shan* (Talki gorge, 1800–2400 m, Aug. 16, 1877—A. Reg., typus!).

General distribution: endemic.

Note. In appearance, this species closely resembles *A. petraeum* Kar. et Kir., i.e., form of inflorescence, colour of perianth, structure of bulbs and scabrousness of leaf blades and sheaths; it differs sharply (like other onions closely related to *A. petraeum*) in short beak of spathe (1.5 mm long), absence of bracts and enlarged filaments of stamens (filaments enlarged for 3/4 their length in inner stamens and 1/2 in outer) which are 1/2 length of perianth. (In species closer to *A. petraeum*, beak of spathe up to 1–1.5 cm long, filaments not enlarged at all and nearly twice longer than perianth.) In these characteristics, *A. megalobulbon* approaches Kashgar onion *A. pevtzovii* Prokh. But in the latter, unlike in *A. megalobulbon*, enlarged parts of filaments above connate part adnate to perianth to form a rather low corona; additionally, its leaves emerge from the scape almost at the same level while leaf sheaths are evidently underground.

43. **A. petraeum** Kar. et Kir. in Bull. Soc. natur. Moscou 15, 3 (1842) 512; Vved. in Fl. SSSR, 4 (1935) 183; Fl. Kirgiz. 3 (1951) 73; Fl. Kazakhst. 2 (1958) 164; Vved. in Opred. rast. Sr. Azii [Key to Plants of Mid. Asia], 2 (1971) 65. —*A. globosum* var. *ochroleucum* Rgl. in Acta Horti Petrop. 3, 2 (1875) 199; Danguy in Bull. Mus. nat. hist. natur. 20, 3 (1914) 139. —*A. saxatile* auct. non M.B.: Vved. in Fl. SSSR, 4 (1935) 184, p.p., quoad pl. As. Centr.

Described from Kazakhstan (Jung. Alatau). Type in Moscow (MW). Plate II, fig. 4.

Rocky and rubble slopes and rocks; from foothills to upper part of middle mountain belt, very rare in subalpine belt.

IIA. Junggar: *Cis-Alt.* (Kandagatai river, Sept. 14, 1876—Pot.); *Jung. Alatau* (ascent to Kuzyun' pass, rocky sites, Aug. 2, 1908—B. Fedtsch.); *Tien Shan* (Talki river, July 18; Sairam lake, July—1877, A. Reg.; Khanakhai brook, 900–1200 m, June 15; Ili river, 1500–2100 m, June 16; Sairam lake, 2100–2700 m, July 29; Pilyuchi, 900–1200 m, July—1878, A. Reg.; Sairam lake, July 19, 1878—Fet.; Borgaty, 1800 m, July 4 and 1500–1800 m, July 5; Aryslyn river, 2400–2700 m, July 10, 1800–2400 m, July 14 and July 19; Turgun-Tsagan, July—1879, A. Reg.; Urumchi region, Tasenku river, Biangou locality, rocky slopes, spruce belt, Sept. 24, 1929—Pop.; 20 km south of Nyutsyuan'tsz, 2500 m, No. 197, July 8; Danu, among rocks, No. 1428, July 16—1957, Kuan; Manas river basin, ulan-Usu river valley 1–2 km before confluence with Koisu, forest belt, bushy steppe along south. rocky slope, July 17; same locality, Koisu river valley at point of entering Ulan-Usu, right flank of valley, in rock crevice, July 19; same locality, Ulan-Usu river valley at confluence with Dzhartas river, subalpine belt, along south. rocky slope, July 19—1957, Yun. et al.; south of Shichan, Kizilzangi village, among rocks, No. 926, July 23; along road to Tyan'chi from Fukan, No. 4274, Sept. 18, 1957—Kuan; "Bords du Sairam-Nor, montagnes, July 23, 1895, Chaff."—Danguy, l.c.); *Dzhark.* (Suidun, July 8, 1877—A. Reg.).

General distribution: Jung.-Tarb., Nor. Tien Shan.

Note. In appearance this species is very close to *A. tianschanicum* Rupr. (the 2 species are often confused during identification). Differs from *A. tianschanicum* in rigid and opaque outer bulb tunic, leaves with scabrous sheaths and leaf blades flat, costate, marginally serrulate, with projecting veins; perianth lobes with short recurved cusp. (Outer tunic of bulbs in *A. tianschanicum* soft, lustrous, leaves with glabrous sheaths and cylindrical or semicylindrical unribbed, glabrous blades, sometimes diffusely serrulate only at base.) In both species, perianth may be albescent, yellowish or pinkish but plants with pinkish and pink perianth lobes are found among *A. tianschanicum* more often than among *A. petraeum.* Both species are closely related to *A. globosum* M.B. ex Redouté.

44. **A. pevtzovii** Prokh. in Izv. Glavn. bot. sada SSSR, 29, 5–6 (1930) 561. —Ic.: l.c. fig. 3.

Described from Kashgar. Type in Leningrad.

Loessial slopes in lower hill belt.

IB. Kashgar: *South.* (nor. Kuen'lun' foothills, near Kok-Yar village, loessial dry clay, 1350 m, 16/28 VI, 1889—Rob., typus!).

General distribution: endemic.

Note. The taxonomic position of *A. pevtzovii* is not quite clear. In appearance, particularly in bulb structure, it is similar to *A. tianschanicum* Rupr. and *A. petraeum* Kar. et Kir. but differs in several characters (low foliation of scape, very short and evidently underground leaf sheaths, greatly enlarged filaments of inner stamens, stamens shorter than perianth, short beak of spathe and absence of bracts).

Ya.I. Prokhanov, the author of the description, compared this species with *A. schoenoprasoides* Rgl. with which it is similar in structure of filaments of stamens, and with *A. globosum* M.B. ex Redouté and *A. lilacinum* Royle. *A. pevtzovii*, however, exhibits no distinct affinity with any of these species. (See also note under *A. megalobulbon* Rgl.)

45. **A. platyspathum** Schrenk in Fisch. et Mey. Enum. pl. nov. 1 (1841) 7; Rgl. in Acta Horti Petrop. 3, 2 (1875) 135; Regel in Izv. Obshch. lyubit. estestvozn. antrop. etnogr. 21, 2 (1876) 73; Rgl. in Acta Horti Petrop. 10, 1

(1887) 330; Danguy in Bull. Mus. nat. hist. natur. 14 (1908) 132; Vved. in fl. SSSR, 4 (1935) 175; Poljak. in Bot. mat. (Leningrad) 12 (1950) 68; Fl. Kirgiz. 3 (1951) 65; Fl. Kazakhst. 2 (1958) 159; Fl. Tadzh. 2 (1963) 309; Ikonn. Opred. rast. Pamira [Key to Plants of Pamir] (1963) 88; Vved. in Opred. rast. Sr. Azii [Key to Plants of Mid. Asia], 2 (1971) 62. —*A. amblyophyllum* Kar. et Kir. in Bull. Soc. natur. Moscou, 15, 3 (1842) 510; Rgl. l.c. (1875) 127; Poljak. l.c. 68; Fl. Kazakhst. 2 (1968) 159. —**Ic.:** Regel, l.c. tab. 12, figs. 4–6; Fl. Kirgiz. 3, Plate 14, fig. 1; Fl. Kazakhst. 2, Plate 13, fig. 10.

Described from Kazakhstan (Jung. Alatau). Type in Leningrad.

Rocky, rubble and melkozem slopes, subalpine meadows and short-grass meadows, alpine spruce forests, low (rarely), middle and upper mountain belts.

IB. Kashgar: *Nor.* (Kucha, No. 10059, July 26, 1959—Lee et al. (A.R. Lee (1959)); *West.* (Sulu-Sakal valley 25 km east of Irkeshtam, July 26, 1935—Olsuf'ev).

IIA. Junggar: *Cis-Alt.* (Kran river, July 24, 1903—Gr.-Grzh.); *Tarb.* (Saur mountain range, Karagaitu river valley, right bank Bain-Tsagan creek valley, subalpine meadow, June 23, 1957—Yun. et al.; nor. of Dachen, 2250 m, No. 1575, Aug. 13, 1957—Kuan); *Jung. Alatau* (Kassan pass, 2700–3000 m, Aug. 11, 1878—A. Reg.; Toli district, Bartok-Arba-Kezen', 2450 m, No. 1202, Aug. 6; mountains in Toli district, on slope, No. 2599, Aug. 7; 15 km nor.-west of Ven'tsyuan' [Arasan], No. 1634, Aug. 29—1957, Kuan); *Tien Shan* (Dzhagastai pass, 2400–2700 m, June 20 and 2400–3000 m; Talki brook near Sairam lake, 3000 m, July 18 and 22; Kyzemchek near Sairam lake, July 31; Muzart pass, 3000–3450 m, Aug. 18—1877, A. Reg.; Khanakhai mountain, 1500–2100 m, May 16; Sumbe pass, 2700–3000 m, June 22; Bogdo mountain, 2100–2400 m, July 24 and 3000 m, July—1878, A. Reg.; Barkhantau, June 5; Muzart pass, Aug. 3; Yuldus, Sept.—1878, Fet.; Nilki on Kash river, 2100 m, June 8; south. tributary of Kara-Gola at pass to Nilki, 2700 m, June 16; between Dzhirgalan and Aryslyn, 2400 m, July 8—1879, A. Reg.; Oberstes Dschanart Tal, June 14–17, 1903—Merzb.; Plateau eines Berges im oberen Kok-su Gebiet, ca. 3400 m, July 8–10, 1907—Merzb.; Danu river valley, south of Danu town, No. 383, July 21; Balun'tai nor. of Bagrashkul' lake, Aug. 6; 20 km nor. of Ulastai, 2300 m, No. 3830, Aug. 28; between Nilki and Ulastai, 1500 m, No. 3915, Aug. 29—1957, Kuan; Manas river basin, Danu-Gol river valley, 1–2 km beyond small opening to Danu pass, short-grass meadow near base of rock, July 21, 1957—Yun. et al.; Narat mountain range, Ardyn-Daban pass from upper Kunges to Yuldus basin, subalpine forb meadow along gullies, Aug 8; Muzart river valley, right bank creek valley of Tunu-Daban near Oi-Terek area, 2900 m, alpine spruce forest, Sept. 7—1958, Yun. and I.-F. Yuan'; from Bortu to timber works at Khomote, on slope, 2520 m, No. 7078, Aug. 4; Narat hill in Kunges, 2300 m, No. 6549, Aug. 7, 1958—Lee and Chu).

IIIC. Pamir (Charlysh river basin, Ulug-Tuz gorge, in meadow among junipers, June 23, 1909—Divn.; "Rochers, vallee de Tor-Bachi, 3800 m, July 31, 1906, Lacoste"—Danguy, l.c.).

General distribution: Jung.-Tarb., Nor. and Cent. Tien Shan, East. Pam.; Mid. Asia (Zeravshan mountain range), West. Sib. (Altay: Narymsk mountain range).

46. A. platystylum Rgl. in Acta Horti Petrop. 10, 1 (1887) 328; Paulsen in Hedin, S. Tibet, 6, 3 (1922) 89. —*A. platyspathum β. falcatum* Rgl. in Acta Horti Petrop. 6 (1879) 135. —**Ic.:** Rgl. l.c. 1887, tab. 11, fig. 2.

Described from Tibet. Type in Leningrad. Plate II, fig. 7.

Alpine meadows, along pebble and silty sites in river valleys, wet clayey slopes, rocky placers and rocks in alpine belt.

IIIA. Qinghai: *Nanshan* (South Tetung mountain range, along Rangkhta-Golu river, right tributary of Tetung, Aug. 8, 1872—Przew.; Nanshan, in silty sites, 3300–3600 m, July 18, 1879—Przew.; South Kuku-Nor mountain range, in silty sites, 3150—3450 m, June 7, 1880—Przew.; nor. slope of Humboldt mountain range, around Ukan-Bulak area, alpine meadow, 3900 m, June 30, 1894—Rob.; nor. bank of Kuku-Nor lake, Baga-Ulan river, Aug. 3, 1890—Gr.-Grzh.; pass through Altyntag mountain range 24 km from Aksai settlement, Aug. 2, 1958—Petr.).

IIIB. Tibet: *Chang Tang* (nor. slope of Przewalsky mountain range, Achik-Kul', along rocks, Aug. 18, 1890—Rob.; "Mandarlik, 3437 m, medio July 1900, Hedin"—Paulsen, l.c.); *South.* ("S.W. Tibet, height above source of Tsangpo, 5015 m, July 13, 1909, Hedin"—Paulsen, l.c.); *Weitzan* (Konchunchu river, 3900-4200 m, July 1; along By-Dzhun river, 3900 m, July 7—1884, Przew.; Dzhagyn-Gol river, pebble bed, 4110 m, July 23, 1884—Przew., lectotypus!; nor. bank of Russkoe lake, Alyk-Norsk valley, clayey slopes, 4050 m, June 26; left bank of Dzhagyn-Gol river, wet clayey soil, 4080 m, July 3; Yangtze river-basin, upper Khichu river in Nyamtso district, rock placers, 4200 m, July 11, 4350 m, July 13—1900, Lad.; nor. slope of Burkhan-Budda mountain range, Khatu gorge, 3300–4500 m, July 11, 1901—Lad.).

General distribution: endemic.

47. **A. setifolium** Schrenk in Fisch. et Mey. Enum. pl. nov. 1 (1841) 6; Rgl. in Acta Horti Petrop. 3, 2 (1875) 109; Regel in Izv. Obshch. lyubit. estestvozn. antrop. etnogr. 21, 2 (1876) 63; Rgl. in Acta Horti Petrop. 10, 1 (1887) 320; Vved. in Fl. SSSR, 4 (1935) 178; Fl. Kazakhst. 2 (1958) 161; Vved. in Opred. rast. Sr. Azii [Key to Plants of Mid. Asia], 2 (1971) 64. — Ic.: Regel, l.c. 1876, tab. 11, fig. 1–3, Fl. Kazakhst. 2, Plate 12, fig. 4.

Described from Kazakhstan (Jung. Alatau: Labasy hills). Type in Leningrad.

Rocky and rubble slopes in foothills.

IIA. Junggar: *Jung. Alatau* (2 km nor. of Laofynkou village, in Toli district, 700 m, No. 173, Aug. 16, 1957—Kuan); *Dzhark.* (Bayandai, 600 m, May 5, 1878—A. Reg.).

General distribution: Jung.-Tarb. (Jung. Alatau), Nor. Tien Shan.

48. **A. subtilissimum** Ledeb. Fl. alt. 2 (1830) 22; Rgl. in Acta Horti Petrop. 3, 2 (1875) 103 and 10, 1 (1887) 318; Kryl. Fl. Zap. Sib. 3 (1929) 612; Vved. in Fl. SSSR, 4 (1935) 178; Fl. Kazakhst. 2 (1958) 161; Vved. in Opred. rast. Sr. Azii [Key to Plants of Mid. Asia], 2 (1971) 64.

Described from West. Siberia (Irtysh river). Type in Leningrad.

Rocky slopes in foothills.

IIA. Junggar: *Jung. Gobi* (nor. Kotbukha [Mukurtai area] on rocky slopes, Aug. 10, 1876—Pot.; near Kyup spring [south. trail of Baitak-Bogdo], Aug. 8, 1898—Klem.).

General distribution: Fore Balkh. (Zaisan basin), Jung.-Tarb.; West. Sib. (southeast).

49. **A. tianschanicum** Rupr. in Mém. Acad. Sci. St.-Pétersb. (Sci. Phys. Math.), ser. 7, 14, 4 (1869) 33; Vved. in Fl. SSSR, 4 (1935) 182; Fl. Kirgiz. 3

(1951) 70; Fl. Kazakhst. 2 (1958) 163; Fl. Tadzh. 2 (1963) 314; Ikonn. Opred. rast. Pamira [Key to Plants of Pamir] (1963) 87; Vved. in Opred. rast. Sr. Azii [Key to Plants of Mid. Asia], 2 (1971) 65. —*A. hymenorrhizum β. tianschanicum* (Rupr.) Rgl. in Acta Horti Petrop. 3, 2 (1875) 132. —*A. globosum* var. *albidum* Rgl. l.c. 10, 1 (1887) 352. —*A. globosum* auct. non M.B. ex Redouté: Danguy in Bull. Mus. nat. hist. natur. 20 (1914) 139. — Ic.: Fl. Kirgiz. 3, Plate 11, fig. 2.

Described from Kirghizstan (Cent. Tien Shan: Dzhaman-Daban). Type in Leningrad. Plate II, fig. 2; map 1.

Rocky slopes and rocks, predominantly in alpine belt, rarely in middle and very rarely in lower mountain belt.

IB. Kashgar: *Nor.* (Sogdan-Tau mountain range, 10–12 km south-east of Akchit water monitoring station on Kokshaal river, Sept. 19, 1958—Yun. and I.-F. Yuan'); *West.* (Sandal pass, Aug. 11, 1890—Grombch.; upper Tiznaf river, 25–27 km beyond Kyude settlement along Tibet highway to Saryk-Daban pass, alpine wormwood desert, 3900 m, June 1; same locality, 35 km beyond Kyude settlement along Tibet highway to Saryk-Daban pass, near Pobedy bridge, parts of alpine steppe among rocks, 3900 m, June 1; Akkez-Daban pass 110 km south of Kargalyk along Tibet highway, hilly steppe, June 5; nor. slope of Kingtau mountain range 2 km nor. of Kosh-Kulak settlement, steppe belt, along gorge, June 10—1959, Yun. and I.-F. Yuan'; 50 km nor. of Baikurt, 3100 m, No. 09738, June 20, 1959—Lee et al. (A.R. Lee (1959)).

IIA. Junggar: *Jung. Alatau* (dwarf juniper groves on Borotala river, 1500–1800 m, Sept. 5, 1880—A. Reg.); *Tien Shan* (Sairam lake, 2100 m, June 21; south-east of Sairam lake, July 23; Talki, July; Dzhagastai, Aug. 8; upper Muzart river below crossing, 2700 m, Aug. 19—1877, A. Reg.; Bogdo mountain, 2400–2700 m, July 24; Sairam lake, Kyzemchek, 2010 m, July; upper Khorgos, 2700 m, Aug. 11; Borborogusun, 1200–1600 m, Aug. 25—1878, A. Reg.; Urtaksary, July 20, 1878—Fet.; Aryslyn, 2400 m, July 8, 1879—A. Reg.; 10–12 km nor. of Shuvutin-Daban pass from Sairam-Nur basin to Borotala, along limestones on flank of Urtaksary river valley, Aug. 18, 1957.-Yun. et al.; 25 km nor.-west and beyond Balinte settlement along Khanga river valley from Karashar to B. Yuldus, among boulders on old moraine, Aug. 1; intermontane B. Yuldus basin 4–5 km south-west of Bain-Bulak settlement, along southern steep limestones slopes, Aug. 10; Muzart river valley, Chokarna area between Kurtan and Oi-Terek, Muzart river floodplain among boulders on alluvium, Sept. 7; right bank Tunu-Daban creek valley near Oi-Terek area, common Russian thistle steppised desert on moraine, 2350 m, Sept. 11—1958, Yun. and I.-F. Yuan'; Khetszin, Logotou region, in rock crevice, No. 6271, Aug. 1; hill road from Bortu to timber plant in Khomote, 1625 m, No. 6955, Aug. 3; 2 km south-west of Bain-Bulak in Khotun-Sumbule, 2650 m, No. 6426, Aug. 10—1958, Lee and Chu (A.R. Lee (1959)); "Environs du Sairam-Nor, montagnes, July 23, 1895, Chaff."—Danguy, l.c.).

IIIB. Tibet: *Chang Tang* (Altyn-Tag mountain range near Chuka-Daban pass, on rocks, 3600 m, 1890—Rob.; Keriya river-basin 18 km south of Polur, steppised wormwood forest on melkozem slope, May 12, 1959—Yun. and I.-F. Yuan').

IIIC. Pamir (Tagdumbash-Pamir, Sarykol mountain range, Pistan gorge, in rocky localities, 3900–4200 m, July 15; same locality along Kara-Chukur river, on rocks, July 18—1901, Alekseenko; Pasrabat river valley, Toili-Bulun area, along clayey-rocky descent, Aug. 2, 1909—Divn.; Muztag-Ata foothill, along rocky placers [undated], Divn.; upper Arialykdarya, 3200 m, July 7; Posarbat area, 2500–3000 m, Aug.-Sept.— 1941, Serp.; Bulunkul' settlement near confluence of 2 sources of Gëzdarya river, grassy winter fat groves, along arid gullies on lacustrine pebble bed terrace, June 12, 1959—Yun.).

General distribution: Jung.-Tarb. (Jung. Alatau), Cent. Tien Shan, East. Pam.; Mid. Asia (Alay mountain range).

Note. Several cited specimens of this species differ from the type specimen in much smaller plant size and lax, few-flowered inflorescence with very short spathe beak. These specimens probably belong to a distinct variety.

This species is similar in appearance to A. *petraeum* Kar. et Kir. (see note under this species).

Section **Petroprasum** F. Herm.

50. **A. obliquum** L. Sp. pl. (1753) 296; Rgl. in Acta Horti Petrop. 3, 2 (1875) 126 and 10, 1 (1887) 327; Kryl. Fl. Zap. Sib. 3 (1929) 614; Vved. in Fl. SSSR, 4 (1935) 175; Fl. Kirgiz. 3 (1951) 65; Fl. Kazakhst. 2 (1958) 158; Vved. in Opred. rast. Sr. Azii [Key to Plants of Mid. Asia], 2 (1971) 62.

Described from Siberia. Type in London (Linn.). Plate III, fig. 3.

Hilly forest meadows in lower and middle mountain belts.

IIA. **Junggar:** *Tien Shan* (Dzhagastai, 1500–2100 m, Aug. 9, 1877—A. Reg.; Khanakhai, June 16, 1878—A. Reg.; Kindikun gorge, 1800 m, July 6 [1878]—Fet.).

General distribution: Jung.-Tarb., Cent. Tien Shan; Europe (Middle Europe and south-east. European part of USSR), Mid. Asia (Fergana mountain range), West. Sib. (south., including Altay), East. Sib. (Ang.-Sayan.).

Section **Schoenoprasum** Koch

51. **A. herderianum** Rgl. in Acta Horti Petrop. 10, 1 (1887) 324; Forbes and Hemsley, Index Fl. Sin. 3 (1905) 122. —**Ic.:** Rgl. l.c. tab. VIII, fig. 2.

Described from specimens grown from bulbs collected in North-west China (Gansu). Type in Leningrad.

IIIA. **Qinghai:** *Nanshan* (nor. bank of Kuku-Nor lake, Baga-Ulan river, July 3, 1890—Gr.-Grzh.).

General distribution: China (North-west).

52. **A. juldusicolum** Rgl. in Acta Horti Petrop. 6, 2 (1880) 523 and 10, 1 (1887) 327.

Described from Junggar (Tien Shan). Type in Leningrad.

IIA. **Junggar:** *Tien Shan* (Yuldus, Sept. 1878—Fet., typus!).

General distribution: endemic.

53. **A. karelinii** Poljak. in Bot. mat. (Leningrad) 12 (1950) 70; Fl. Kirgiz. 3 (1951) 81; Fl. Kazakhst. 2 (1958) 168; Vved. in Opred. rast. Sr. Azii [Key to Plants of Mid. Asia], 2 (1971) 68. —*A. schoenoprasum* f. *scaberrimum* Kar. et Kir. in Bull. Soc. natur. Moscou, 15, 3 (1842) 507. —*A. schoenoprasum* δ.

scaberrimum (Kar. et Kir.) Rgl. in Izv. Obshch. lyubit. estestvozn. antrop. etnogr. 21, 2 (1876) 45, p.p.; Rgl. in Acta Horti Petrop. 10, 1 (1887) 306.— *A. schoenoprasum* auct. non L.: Sap. Mong. Alt. (1911) 388, p.p., quoad pl. ex lac. Dain-Gol.; Vved. in Fl. SSSR, 4 (1935) 190, p.p. —**Ic.:** Fl. Kazakhst. 2, Plate 12, fig. 5.

Described from Kazakhstan (Jung. Alatau). Type in Leningrad.

Moist grasslands and meadows in alpine belt.

IA. Mongolia: *Mong. Alt.* (Boroborgosun river, Saksai tributary, July 2, 1903—Gr.-Grzh.; Tsagan-Gol, upper camp near cold spring, July 3, 1905—Sap.; Dain-Gol lake, on slopes, July 28, 1909—Sap.; Tamchi somon, Bus-Khairkhan mountain range, gully between steep slopes, July 17; Khargatiin-Daba, alpine meadow, July 23; Bulugun somon, Kharagaitu-Daba pass, alpine meadow, July 24; Bulugun river-basin, Kharagaitu-Khutul' crossing, alpine meadow and rubble placers, July 24; same locality, upper course of Ketsu-Sairin-Gol river, alpine meadow, July 26; Adzhi-Bogdo mountain range, upper third of Mainigtu-Ama creek valley, changeover from alpine steppes to talus coenoses, Aug. 7—1947, Yun.; upper course Kobdo river, Donyagiin-Khara-Nur lake, east. bank of nor. extremity along road to Dayan-Nur, meadow steppe along west. slope, July 7, 1971—Grub. et al.).

IIA. Junggar: *Cis-Alt.* (Kandagatai river, Sept. 18, 1876—Pot.); *Tien Shan* (Yuldus, Sept. 1878—Fet.; Aryslyn, 2700 m, July 11, 1879—A. Reg.).

General distribution: Jung.-Tarb., Nor. and Cent. Tien Shan; Mid. Asia (West. Tien Shan).

Note. Very closely related to *A. schoenoprasum* L., differing in serrulate-scabrous scape and leaf sheaths and blades (along projecting veins). Flowers of *A. karelinii* are white, pinkish or pink, sometimes coral-red (as in *A. schoenoprasum*). The species requires further investigation.

54. **A. oliganthum** Kar. et Kir. in Bull. Soc. natur. Moscou, 14 (1841) 856; Rgl. in Acta Horti Petrop. 3, 2 (1875) 115 and 10, 1 (1887) 323; Vved. in Fl. SSSR 4 (1935) 195; Grub. Konsp. fl. MNR (1955) 94; Fl. Kazakhst. 2 (1958) 169; Vved. in Opred. rast. Sr. Azii [Key to Plants of Mid. Asia], 2 (1971) 69. —*A. stenophyllum* Schrenk in Bull. Phys.-Math. Acad. Sci. (Pétersb.) 3 (1845) 210; Rgl. l.c. (1875) 93 and l.c. (1887) 315; Kryl. Fl. Zap. Sib. 3 (1929) 613; Grub. l.c. 94.

Described from Kazakhstan (Fore Balkhash). Type in Leningrad.

Moist meadows, marsh borders, plains and foothills.

IA. Mongolia: *Khobd.* (Achit-Nur lake, marshy region between Bukhu-Muren and Khubusu-Gol rivers 7–8 km east-north-east of Bukhu-Muren somon, solonetz sedge meadow, July 15, 1971—Grub. et al.); "*Mong. Alt.*"—Grub. l.c.; *Depr. Lakes* ("Ubsa-Nur, Kaady river"—Grub. l.c.).

IB. Kashgar: *East.* (Turfan: north-east of Toksun, swamp, 300 m, No. 7324, June 19, 1958—Lee and Chu).

IIA. Junggar: *Tien Shan* (Urumchi region, path between Tien Shan and Bogdo-Ol, near Saepu village and lake, Sept. 5, 1929—Pop.); *Jung. gobi* (Bodonchi river floodplain 2–3 km south of Bodonchi-Khure, solonchak meadow along tributaries, 1947—Yun.).

General distribution: Fore Balkh.; Mid. Asia (Turkestan mountain range), West. Sib. (south.).

68

55. **A. schoenoprasum** L. Sp. pl. (1753) 301; Rgl. in Acta Horti Petrop. 3, 2 (1875) 77 and 10, 1 (1887) 306, pro max. p.; Sap. Mong. Alt. (1911) 388; Kryl. Fl. Zap. Sib. 3 (1929) 606; Vved. in Fl. SSSR, 4 (1935) 190, p.p.; Kitag. Lin. Fl. Mansh. (1939) 132; Grub. Konsp. fl. MNR (1955) 94; Vved. in Opred. rast. Sr. Azii [Key to Plants of Mid. Asia], 2 (1971) 69.

Described from Europe and Siberia. Type in London (Linn.).

Humid and marshy meadows in river valleys.

IA. **Mongolia:** *Khobd.* (near Altyn-Khatysyn, on sand, June 18, 1879—Pot.); *Mong. Alt.* (Iter ad Kobdo, June 20, 1870—Kalning; Saksai river valley, Kobdo tributary, July 3, 1908—Utkin; "Dzhyumaly; M. Kairta river; Urmogaity river"—Sap. l.c.); *Depr. Lakes* (Dzergin river, saline meadows along banks of river valley, Aug. 16, 1930—Bar.).

IIA. **Junggar:** *Jung. Gobi* (Nom, June 20, 1877—Pot.).

General distribution: Jung.-Tarb.; Arct., Europe, Caucasus, West. Sib., East. Sib., Far East, Nor. Mong., China (Dunbei), Japan, Nor. Amer.

Note. Plants cited by V.V. Sapozhnikov (l.c.) for Mong. Altay perhaps belong to the related and widely distributed *A. karelinii* Poljak.

Section **Annuloprason** Egor.[4]

56. **A. atrosanguineum** Kar. et Kir. in Bull. Soc. natur. Moscou, 15, 3 (May 23, 1842) 508; Poljak. in Bot. mat. (Leningrad) 12 (1950) 69. —*A. atrosanguineum* Schrenk in Bull. Sci. Acad. Sci. St.-Pétersb. 10, 23 (July 18. 1842) 355; Rgl. in Acta Horti Petrop. 3, 2 (1875) 83; Regel in Izv. Obshch. lyubit. estestvozn. antrop. etnogr. 21, 2 (1876) 46; Fl. Kirgiz. 3 (1951) 78; Fl. Kazakhst. 2 (1958) 167; Vved. in Fl. Tadzh. 2 (1963) 317; Ikonn. Opred. rast. Pamira [Key to Plants of Pamir] (1963) 88; Vved. in Opred. rast. Sr. Azii [Key to Plants of Mid. Asia], 2 (1971) 67. —*A. monadelphum* γ. *atrosanguineum* (Schrenk) Rgl. in Acta Horti Petrop. 10, 1 (1887) 309. —*A. monadelphum* auct. non Less. ex Kunth: Vved. in Fl. SSSR, 4 (1935) 189, quoad syn. *A. atrosanguineum* Schrenk. —Ic.: Regel, l.c. 1876, tab. VIII, figs. 4–5; Fl. Kazakhst. 2, Plate 12, fig. 9.

Described from Kazakhstan (Jung. Alatau). Type in Leningrad.

Moist slopes in subalpine and alpine hill belts.

IIA. **Junggar:** *Jung. Alatau* (upper Khorgos, 2700–3000 m [May 1878]—A. Reg.; Kassan pass, 2700–3000 m, June 10, 1878—A. Reg.; 30 km west of Ven'tsyuan' [Arasan], 2640 m, No. 2054, Aug. 5, 1957—Kuan); *Tien Shan* (Sairam lake, Talki brook, July 19, 1877—A. Reg.; Pilyuchi pass, 2100–2400 m, June 26, 1878; Kok-Kamyr

---

[4]Section **Annuloprason** Egor., sect. nova. —Bulbi rhizomati adnati anguste cylindrico-conici. Filamenta 1/3–1/2 tepalorum longitudinis attingentia perigonio per dimidium longitudinis filamentorum circiter adnata, superne a perigonio libera in annulum connata. Stylus stigmate 3-lobato.

Typus: *A. fedtschenkoanum* Rgl.

mountains, 3000 m, July 25, 1878—A. Reg.; Urtaksary, July 20, 1878—Fet.; bei Manas, July 5–15, 1908—Merzb.).

General distribution: Jung.-Tarb., Nor. and Cent. Tien Shan, East. Pam., Mid. Asia (West. Tien Shan, Pam.-Alay).

Note. See note under *A. fedtschenkoanum* Rgl.

57. **A. chalcophengos** Airy-Shaw in Notes Roy. Bot. Gard. Edinb. 16 (1931) 137. —*A. monadelphum v. tibeticum* Rgl. in Acta Horti Petrop. 10, 1 (1887) 311; Forbes and Hemsley, Index Fl. Sin. 3 (1905) 123; Rehder in J. Arn. Arb. 14 (1933) 5. —*A. monadelphum* var. *fedtschenkoanum* (Rgl.) Rgl. l.c. 308, p. min. p., quoad pl. tibet. —*A. semenovii* auct. non Rgl.: Hemsley and Pearson in Peterm. Mitt. 28 (1900) 375; Hemsley in J. Linn. Soc. (London) Bot. 35 (1902) 199; Paulsen in Hedin, S. Tibet, 6, 3 (1922) 89.

Described from South-west China (Sichuan). Type in London (K). Plate III, Fig. 4.

Alpine belt.

IIIA. **Qinghai:** *Nanshan* (toward Khsan river sources, May 23, 1886—Pot.; South Kuku-Nor mountain range, Karagatu area, 3900 m, May 4, 1895—Rob.); *Amdo* (Dzhakhar-Dzhargyn mountain range, 3900 m, June 26, 1880—Przew.).

IIIB. **Tibet:** *Weitzan* (south. slopes of mountain range between Huang He and Yantszytszyan rivers [descent into watershed between Huang He and Murusu (Yangtze)], in rocky places, June 12; along Daochu river, 3900 m, July 12—1884, Przew.; Yantszytszyan basin, Nyamtso district along Khichu river, 3200 m, July 12, 1900—Lad.; Yantszytszyan basin, Namsun pass, 4500 m and Lamalunla pass, 4380 m, June 30, 1901—Lad.; "Eastern Tibet: alpine region between Radja and Jurap ranges, Rock"—Rehder, l.c.; "Northern Tibet, camp 31, 4616 m, Sept. 21, 1896, Hedin"— Paulsen, l.c.; "Widely distributed: 88–96°, 35°10'–35°20' 4380–5100 m"—Hemsley, l.c.).

General distribution: China (North-west, South-west).

58. **A. fedtschenkoanum** Rgl. in Acta Horti Petrop. 3, 2 (1875) 82; Regel in Izv. Obshch. lyubit. estestvozn. antrop. etnogr. 21, 2 (1876) 45; Fl. Kirgiz. 3 (1951) 78; Fl. Tadzh. 2 (1963) 316; Vved. in Opred. rast. Sr. Azii [Key to Plants of Mid. Asia], 2 (1971) 68. —*A. monadelphum β. fedtschenkoanum* (Rgl.) Rgl. in Acta Horti Petrop. 10, 1 (1887) 311, pro max. p., excl. pl. tibet. —*A. monadelphum* auct. non Less. ex Kunth: Vved. in Fl. SSSR, 4 (1935) 189, p.p., quoad syn. *A. fedtschenkoanum* Rgl. —Ic.: Regel, l.c. 1876, tab. 7, figs. 4–6; Fl. Tadzh. 2, Plate 57, fig. 3.

Described from Mid. Asia (Zeravshan river valley). Type in Leningrad, isotype in London (K). Plate III, fig. 7.

Humid meadows and slopes in upper hill belt.

IIA. **Junggar:** *Tien Shan* (Sairam lake, Talki brook, July 19, 1877—A. Reg.).

General distribution: Fore Asia (Afghanistan, Pakistan), Mid. Asia (West. Tien Shan, Pam.-Alay).

Note. Closely related to *A. atrosanguineum* Kar. et Kir., differing from it in colour and form of perianth lobes: in *A. fedtschenkoanum*, these are yellow (later erubescent, turning yellow again), oblong-lanceolate or narrowly lanceolate, usually with attenuated, acute or subacute tips; in *A. atrosanguineum*, these are atropurpureous, oblong-ovate, rounded (not attenuated) at tip. At anthesis (or in herbarium), perianth even of *A. atrosanguineum* turns yellow and then plant identification should be based only on shape of perianth lobes. According to "Flora Kazakhstana" [Flora of Kazakhstan] and "Flora Tadzhikistana" [Flora of Tajikistan], perianth lobes in *A. atrosanguineum* are acute and often with attenuated (as in *A. fedtschenkoanum*) tips. These reports are evidently erroneous. Type specimens of *A. atrosanguineum* and specimens of the species cited here bear perianth lobes as described above. Regel (l.c. 1875) distinguishes *A. fedtschenkoanum* from *A. atrosanguineum* in the key by undivided (not fibrous) bulb tunic but, in fact, bulbs of *A. fedtschenkoanum* also have a fibrous tunic.

59. **A. monadelphum** Less. ex Kunth, Enum. pl. 4 (1843) 393; Rgl. in Acta Horti Petrop. 3, 2 (1875) 85; p.p., quoad pl. sibir. and 10, 1 (1887) 307, p.p., quoad α. *typicum*, excl. pl. turkest.; Gr.-Grzh. Zap. Mong. 3, 2 (1930) 807; Vved. in Fl. SSSR, 4 (1935) 189, p.p., excl. syn. *A. fedtschenkoanum* Rgl. and *A. atrosanguineum* Kar. et Kir.; Poljak. in Bot. mat. (Leningrad) 12 (1950) 69; Grub. Konsp. fl. MNR (1955) 93.

Described from East. Siberia (West. Sayan). Type in Leningrad.

Dampish grasslands, rocky slopes, brook-banks; alpine belt of mountains and upper forest belt.

IA. Mongolia: *Khobd.* ("Upper Sagliin-Gola"—Grub. l.c.); *Mong. Alt.* ("Valley of Boroborgusun river, tributary of Saksai river, in meadow, July 14, 1903"—Gr.-Grzh. l.c.).

General distribution: East. Sib. (south.), Nor. Mong. (Fore Hubs., Hent.).

60. **A. semenovii** Rgl. in Bull. Soc. natur. Moscou, 41, 1 (1868) 449; Rgl. in Acta Horti Petrop. 3, 2 (1875) 85; Regel in Izv. Obshch lyubit. estestvozn. antrop. etnogr. 21, 2 (1876) 49; Rgl. in Acta Horti Petrop. 10, 1 (1887) 311; Vved. in Fl. SSSR, 4 (1935) 189; Fl. Kirgiz. 3 (1951) 77; Fl. Kazakhst. 2 (1958) 168; Vved. in Opred. rast. Sr. Azii [Key to Plants of Mid. Asia], 2 (1971) 67.—*A. tristylum* Rgl. in Acta Horti Petrop. 10, 1 (1887) 333. —**Ic.:** Regel, l.c. 1876, tab. 8, figs. 4–5.

Described from Kirghizstan (Cent. Tien Shan). Type in Leningrad. Plate III, fig. 2; map 2.

Meadows in middle and upper mountain belts.

IIA. Junggar: *Jung. Alatau* (Chzhaosu district, between Syata and Ven'tsyuan' [Arasan], along boundary, 2000 m, No. 3382, Aug. 13; 20 km south of Ven'tsyuan', 2870 m, No. 1500, Aug. 14; Ven'tsyuan', No. 4647, Aug. 25—1957, Kuan); *Tien Shan* (Muzart river, 1886—Krasnov; M. Yuldus river, 2250–2700 m, May 26, 1877—Przew.; Aktash in Dzhagastai mountains, Aug. 11, 1877—A. Reg.; Urten-Muzart pass, Aug. 2, 1877—Fet.; Yuldus, Sept. 1878—Fet.; Borborogusun, 1800 m, April 27; Taldy upper course, 2400–2700 m, May 17 and 18; Taldy midcourse, 2100 m, May 26; Irenkhabirga mountain range, Taldy river, 2700 m, May 26; Dumbedan and Kum-Daban, nor. slopes of Irenkhabirga mountain range, 2400–2700 m, May 28; Bagaduslung, Dzhin river [Tsaganusu], 2100–2700 m, April 4; Chungur-Daban pass, June 13; Kara-Gol, near crossing to Nilki, 3000 m, June 17; Aryslyn, 2400–2700 m, Aug. 11—1879, A.

Reg.; Tekes river, in meadow by river, 3300 m, June 24, 1893—Rob.; along Danu river, in meadow, 3000 m, No. 2164, July 21; Aksu district, Kentelek mountains, 2800 m, No. 3510, Aug. 14; 3 km south of Yakou, on slope, 2900 m, No. 1682, Aug. 30—1957, Kuan; vicinity Kunges town, 2300 m, No. 6512, Aug. 7, 1958—Lee and Chu); Manas river-basin, upper course of Danu-Gol river near side entrance to Danu pass, moraines, alpine kobresia meadow, July 23, 1957—Yun. et al.; nor. slope of Narat mountain range, foothills descending into Tsanma valley, subalpine forb meadow, Aug. 7; Narat mountain range, south. slope from Dagit pass into Yuldus basin, subalpine meadow along gullies, Aug. 8—1958, Yun. and I.-f. Yuan'; vicinity Kunges town, 2300 m, No. 6512, Aug. 7, 1958—Lee and Chu (A.R. Lee (1959)).

General distribution: Jung.-Tarb. (Jung. Alatau), Cent. Tien Shan.

Note. *A. tristylum* Rgl. l.c. described from East. Tien Shan (Taldy river) is similar to *A. semenovii*, as convincingly demonstrated by a comparison of type and authentic specimens of these species.

A. semenovii differs from the closely related species *A. fedtschenkoanum* Rgl. and *A. atrosanguineum* Kar. et Kir. in flat leaves; because of their longitudinal folding, these leaves may be erroneously treated as fistular in the herbarium. (Attention should be paid to leaf margin which is generally thinned and semitransparent in flat leaves.) *A. semenovii* also differs from the aforesaid species in long (usually (12) 15–17 mm) perianth lobes, often with highly attenuated acute tips.

61. **A. weschnjakowii** Rgl. in Acta Horti Petrop. 6 (1880) 431 and 10, 1 (1887) 342, 293; Danguy in Bull. Mus. nat. hist. natur. 20, 3 (1914) 140; Vved. in Fl. SSSR, 4 (1935) 188; Fl. Kirgiz. 3 (1951) 77; Fl. Kazakhst. 2 (1958) 167; Vved. in Opred. rast. Sr. Azii [Key to Plants of Mid. Asia], 2 (1971) 67. —**Ic.:** Rgl. l.c. 1887, tab. 8, fig. 3; Fl. Kirgiz. 3, Plate 11, fig. 1.

Described from Kazakhstan (Tien Shan: Kegen' river). Type in Leningrad.

Rock steppes and semidesert mountain slopes; lower mountain belt.

IIA. Junggar: *Jung. Alatau* (15 km west of Ven'tsyuan', in steppe, No. 4568, Aug. 21, 1957—Kuan); *Tien Shan* (Borborogusun, 900–1200 m, Aug. 25, 1879—A. Reg.; along road to Sairam-Nur from Borotala through Urtaksary, desert montane steppe, Aug. 18, 1957—Yun. et al.; along Urumchi-Ili highway, No. 2151, Aug. 28, 1957—Kuan; "Koustai, montagnes, terrains secs, pres de Ebi-Nor, July 25, 1895, Chaff."—Danguy, l.c.); *Jung. Gobi* (west.: east of San'tai, on roadside, No. 4738, Aug. 29, 1957—Kuan; 30 km nor. of Sairam-Nur lake, wormwood-onion semidesert in intermontane flat, 1150 m, Aug. 31, 1959—Petr.).

General distribution: Cent. Tien Shan.

Note. This species stands isolated in section Annuloprason because of the equal length of pedicels. In all other species of this section, pedicels of marginal (outer) flowers in the inflorescence are invariably shorter than those of inner flowers.

Section **Narkissoprason** F. Herm.

62. **A. forrestii** Diels in Notes Roy. Bot. Gard. Edinb. 5 (1912) 302; Walker in Contribs U.S. Nat. Herb. 28 (1941) 602.

Described from South-west China. Type in London or Edinburgh.

IA. **Mongolia:** *Alash. Gobi* (around Alashan: "Ho Lan Schan, at edge of woods, 1923, Ching"—Walker, l.c.).

General distribution: China (South-west).

Note. The original author points out that *A. forrestii* is notable for its large flowers and is similar to small specimens of *A. narkissiflorum* Vill. from the Alps but Walker (l.c.) perhaps regarded *A. mongolicum* Rgl. found in Alashan Gobi, also bearing large flowers, as *A. forrestii*.

Section **Cepa** (Mill.) Prokh.

63. **A. galanthum** Kar. et Kir. in Bull. soc. natur. Moscou, 15, 3 (May 23, 1842) 508; Rgl. in Acta Horti Petrop. 3, 2 (1875) 87; Regel in Izv. Obshch. lyubit. estestvozn. antrop. etnogr. 21, 2 (1876) 50; Danguy in Bull. Mus. nat. hist. natur. 20, 3 (1914) 139; Kryl. Fl. Zap. Sib. 3 (1929) 609; Vved. in Fl. SSSR, 4 (1935) 197; Fl. Kirgiz. 3 (1951) 81; Fl. Kazakhst. 2 (1958) 170; Vved. in Opred. rast. Sr. Azii [Key to Plants of Mid. Asia], 2 (1971) 69; Grub. in Bot. zh. 60, 7 (1975) 956. —*A. pseudocepa* Schrenk in Bull. Sci. Acad. Sci. St. Pétersb. 10, 23 (July 18, 1842) 355. —**Ic.:** Regel, l.c. (1876) tab. 8, figs. 6–8.

Described from East. Kazakhstan (Lepsa river). Type in Leningrad.

Rubble slopes, talus; from foothills to middle mountain belt.

IA. **Mongolia:** *Mong. Alt.* (Bulugun river basin, left tributary of Ulistei-Gol, among rocks under gently sloping forest, May 28, 1973—Golubkova and Tsogt).

IIA. **Junggar:** *Jung. Alatau* (on Ili-Urumchi highway 8 km nor. of Bole [Dzhimpan'], on slope, No. 4752, Aug. 29, 1957—Kuan); *Tien Shan* (Ili valley, 42 km east of bridge on Kash river along road to Ziekta, right flank of Ili-Kunges valley, along steep rocky slopes, Aug. 29, 1957—Yun. et al.; "Montagnes entre le Sairam-Nor et l'Ebr-Nor, July 25, 1895, Chaff."—Danguy, l.c.); *Jung. Gobi* (east.: Oshigiin-Ula mountain range, on talus, July 18; same locality, in hills, July 18—1947, Yun.; Baitak-Bogdo-Nuru, Takhiltu-Ula, left creek valley of Ulyastu-Gola gorge, talus on south. slope, 2400 m, Sept. 17 and 18, 1948—Grub.; south.: Urumchi town vicinity, Bogdo-Ola foothills, Sept. 14, 1929—Pop.; Manas river valley, rubble slope, 1150 m, No. 3715, Oct. 2, 1956—Ching); *Zaisan* (along Burchum river, 550 m, No. 2829, Aug. 29; Barbagai, 550 m, No. 2829, Sept. 8—1956, Ching).

General distribution: Aralo-Casp., Fore Balkh., Jung.-Tarb., Nor. and Cent. Tien Shan; West. Sib. (south-east., excluding Altay).

Section **Phyllodolon** (Salisb.) Prokh.

64. **A. altaicum** Pall. Reise, 2 (1773) 737; Vved. in Fl. SSSR, 4 (1935) 196; Grub. Konsp. fl. MNR (1955) 92; Fl. Kazakhst. 2 (1958) 170; Vved. in Opred. rast. Sr Azii [Key to Plants of Mid. Asia], 2 (1971) 69. —*A. fistulosum* auct. non L.: Regel in Izv. Obshch. lyubit. estestvozn. antrop. etnogr. 21, 2 (1876) 51; Rgl. in Acta Horti Petrop. 10, 1 (1887) 313, p.p.;

Sap. Mong. Alt. (1911) 388; Kryl. Fl. Zap. Sib. 3 (1929) 609, p.p. —Ic.: Pall. l.c. tab. R.

Described from altay. Type in Leningrad. Plate II, fig. 1.

Rock screes, rocks, steppe rubble slopes, forest meadows, pebble beds on gorge floor; from lower to upper mountain belts.

IA. Mongolia: *Mong. Alt.* (Upper Kobdos lake, forest meadows, June 27, 1906— Sap.; Khasagtu-Khairkhan, nor. slope of Tsagan-Irmyk-Ul, upper Khuinkerin-Ama, steppe rubble slope, 2600 m, Aug. 23, 1972—Grub. et al.); *Depr. Lakes* (Dzun-Dzhirgalantu mountain range, south-west. slope, Ulyastyn-Gola gorge, 1850–2800 m, June 28, 1971—Grub. et al.); *Gobi-Alt.* (Baga-Bogdo humid canyons, 1800–2400 m, 1925—Chaney; Ikhe-Bogdo mountain range, Bityuten-Ama creek valley, creek valley slopes and alpine zone, Aug. 12, 1927—Simukova; Dundu-Saihan hills, rubble placers along mountain slope, Aug. 17, 1931—Ik.-Gal.; nor. slope of Ikhe-Bogdo mountain range, lower part of Khaptsagaitu creek valley, rubble talus, Sept. 12, 1943—Yun.).

IIA. Junggar: *Cis-Alt.* (Chingil' [Quinhe], in forest, No. 1468, Aug. 7, 1956—Ching).

General distribution: Jung.-Tarb.; West. Sib. (Altay), East. Sib. (south.), Nor. Mong. (Fore Hubs., Hent, Hang.), China (Altay).

Note. In the Mongolian People's Republic, *A. fistulosum* L., closely related to *A. altaicum* and not found as an escape, was formerly cultivated.

Section **Rhizomatosa** Egor.[5]

65. *****A. caespitosum** Stev. ex Bong. et Mey. in Bull. Sci. Acad. Sci. St.-Pétersb. 8 (1841) 341; Rgl. in Acta Horti Petrop. 3, 2 (1875) 159; Regel in Izv. Obshch. lyubit. estestvozn. antrop. etnogr. 21, 2 (1876) 80; Rgl. in Acta Horti Petrop. 10, 1 (1887) 342; Danguy in Bull. Mus. nat. hist. natur. 17, 7 (1911) 557; Kryl. Fl. Zap. Sib. 3 (1929) 622; Vved. in Fl. SSSR, 4 (1935) 174; Fl. Kazakhst. 2 (1958) 158; Vved. in Opred. rast. Sr. Azii [Key to Plants of Mid. Asia], 2 (1971) 62. —Ic.: Regel, l.c. (1876), tab. 14, figs. 10–12.

Described from West. Siberia (Irtysh river—Piketnaya Rybalka). Type in Leningrad.

IIA. Junggar: (Zaisan-Dschungariae orientalis arenosis ad lacum Saissannor, Sievers"—Rgl. l.c., 1887).

Note. Two specimens without exact locality or date of collection or collector are known from Junggar. One bears the label: "Songaria chinensis ad lacum Saisan-Nor". The label on the other reads: "Songaria chinensis, herb. Dr. Kühlewein" (from J. Klinge's herbarium). The former specimen was cited by Regel (l.c., 1887) as collected or brought by Sievers (Sievers in parentheses).

General distribution: Fore Balkh. (Zaisan basin, Balkhash-Alakul' region), West. Sib. (south.-east.: Irtysh), Nor. Mong. (Hang.).

---

[5]Sect. Rhizomatosa Egor., sect. nova. —Planta rhizomate longe repenti; caule ebulboso, basi vaginis albo-hyalinis plus minusve fibroso-laciniatus.

Typus: *A. caespitosum* Siev. ex Bong. et Mey.

Note. The species is notable for long funiform branched rhizome and absence of bulbs, features which justify its classification in a distinct section. In inflorescence and flowers, it is similar in appearance to *A. mongolicum* Rgl. from which (if underground parts are poorly excavated) it may be distinguished by characteristics pointed out in the key.

Subgenus BROMATORRHIZA Ekberg

Section Bromatorrhiza

66. A. fasciculatum Rendle in J. Bot. (London) 44 (1906) 42. Described from Tibet. Type in London (K).

IIIB. Tibet: ("Khambajong, July 1903, Younghusband; ibid. Sept. 1903, Prain"—Rendle, l.c.; "Gyangtse, July-Sept. 1904, Walton"—Rendle, l.c.).
General distribution: Himalayas (east.).

Subgenus ALLIUM

Section Scorodon C. Koch

67. A. caesium Schrenk in Bull. Phys.-Math. Acad. Sci. (St.-Pétersb.), 2 (1844) 113; Regel in Obshch. lyubit. estestvozn. antrop. etnogr. 21, 2 (1876) 40; Rgl. in Acta Horti Petrop; 10, 1 (1887) 304; Vved. in Fl. SSSR, 4 (1935) 222, pro max. p.; Fl. Kirgiz. 3(1951) 85; Fl. Kazakhst. 2 (1958) 178; Fl. Tadzh. 2 (1963) 326; Vved. in Opred. rast. Sr. Azii [Key to Plants of Mid. Asia], 2 (1971) 74. —*A. urceolatum* Rgl. in Acta Horti Petrop. 2 (1873) 405 and 10, 1 (1887) 312; Sap. Mong. Alt. (1911) 387; Gr.-Grzh. Zap. Mong. 3, 2 (1930) 807. —*A. renardii* Rgl. in Acta Horti Petrop. 6 (1880) 521 and 10, 1 (1887) 322; Fl. Kazakhst. 2 (1958) 179. —**Ic.:** Regel, l.c. (1876) tab. 6, figs. 5–7; Fl. Kirgiz. 3, Plate 10.

Described from Kazakhstan (Karaganda). Type in Leningrad.
Steppes, melkozem and rubble slopes.

IIA. Junggar: *Cis-Alt.* ("Val. of Kurtu river, clayey-sandy cliff, July 2, 1903"—Gr.-Grzh. l.c.); *Jung. Gobi* (nor.: "Koksun mountain range, valley of Uzun-Bulak brook, June 16; Irtysh coastal sand, June 18; Kostuk gorge in foothills of south. Altay, June 21, 1903—Gr.-Grzh. l.c.); *Dzhark.* (left bank of Ili south of Kul'dzha, May 27, 1877—A. Reg.); *Balkh.-Alak.* (steppe between Emil and Chigir, June 20, 1905—Obruchev; "Emil' river, steppe"—Sap. l.c.).
General distribution: Aralo-Casp., Fore Balkh., Jung.-Tarb., Nor. and Cen. Tien Shan; Fore Asia (Afghanistan), Mid. Asia (West. Tien Shan, Pam.-Alay), West. Sib. (south., excluding Altay).

68. A. coeruleum Pall. Reise, 2 (1773) 727; Rgl. in Acta Horti Petrop. 10, 1 (1887) 315; Danguy in Bull. Mus. nat. hist. natur. 20, 3 (1914) 139; Kryl. Fl. Zap. Sib. 3 (1929) 611; Vved. in Fl. SSSR, 4 (1935) 221; Fl. Kirgiz, 3

(1951) 85; Fl. Kazakhst. 2 (1958) 178; Grub. in Bot. mat. (Leningrad) 19 (1959) 534; Fl. Tadzh. 2 (1963) 325; Vved. in Opred. rast. Sr. Azii [Key to Plants of Mid. Asia], 2 (1971) 73. —*A. viviparum* Kar. et Kir. in Bull. Soc. natur. Moscou, 14 (1871) 852; Rgl. l.c. 316. —**Ic.:** Pall. l.c. tab. R.

Described from Kazakhstan (Semipalatinsk vicinity). Type in Leningrad. Plate II, fig. 5.

Steppe and meadow-steppe slopes; lower, middle and rarely upper mountain belts.

IA. Mongolia: *Khesi* (Tszyutsyuan', in field, No. 39, July 27, 1956—Ching).

IIA. Junggar: *Tien Shan* (Kunges river upper course, 1200 m, June 24 and 27—1877, Przew.; Dzhagastai, 1200–1500 m, July 1877—A. Reg.; Urtas-Aksu, June 17; Urtaksary river, west of Sairam-Nur lake, July 19, 1878—Fet.; Khanakhai brook, June 15, 1878—A. Reg.; south. tributary of Kara-Gol, 2400–2700 m, May 16; Yuldus, Taldy, 3000–3300 m, May 26; Nilki brook of Kash river, 2100 m, June 8; Borgaty, 1800 m, July 4; lower Aryslyn, 2100–2400 m, July 20—1879, A. Reg.; Syrt von Kara-Bulak und Kara-Dschon, July 30; Kapsalon Tal und neben Tal Kisyl-Sal, auch Plateauhohe von Kara-Dschon, Aug. 2–3—1907, Merzb.; top of a Manas river opening on right, 65 km south of Manas town, 2500 m, June 26, 1954—Mois.; 20 km south-west of Urumchi town, 1030 m, No. 0417, July 19, 1956—Ching; in San'daokhetszy region, on slope, No. 718, June 9; same locality, No. 4342, July 4; 26 km nor. of Nyutsyuan'tsz, shady slope, No. 4537, July 15; south of Shichan, south. slope Nos. 770, 821, July 23; mountains south of Sin'yuan' [Taldy] town, shaded slope, 1900 m, No. 1161, Aug. 22—1957, Kuan; Tien Shan debris cones in Junggar plain 3 km east of Santokhodze on road to Manas, Khorgos river valley, first terrace above meadow along dry irrigation ditch, July 4, 1957—Yun. et al.; "Montagnes calcaires, alt. 1700 m, entre le Turkestan et la Mongolie pres de Gorgosse, July 17, 1895, Chaff."—Danguy, l.c.; "Nor. slope of Merzbacher mountain range, Erdynkho river valley, on high erosion terrace, 1100–1200 m, July 8, 1952, Mois."—Grub. l.c.); *Dzhark.* (Ili bank west of Kul'dzha, May; Pilyuchi near Kul'dzha, July; Talki, July—1887, A. Reg.).

General distribution: Aralo-Casp., Fore Balkh., Jung.-Tarb., Nor. and Cent. Tien Shan; Europe (south-east European part of USSR), Mid. Asia (Pam.-Alay), West. Sib. (south., excluding Altay).

69. **A. delicatulum** Siev. ex Schult. et Schult. f. in Roem. et Schult. Syst. Veg. 7 (1830) 1133; Rgl. in Acta Horti Petrop. 3, 2 (1975) 95; Regel in Izv. Obshch. lyubit. estestvozn. antrop. etnogr. 21, 2 (1876) 54; Kryl. Fl. Zap. Sib. 3 (1929) 610, p.p., excl. syn. *A. viridulum* Ledeb.; Gr.-Grzh. Zap. Mong. 3, 2 (1930) 807; Vved. in Fl. SSSR, 4 (1935) 219; Fl. Kazakhst. 2 (1958) 176; Vved. in Opred. rast. Sr. Azii [Key to Plants of Mid. Asia], 2 (1971) 73. — Ic.: Regel, l.c. 1876, tab. 9, figs. 7–9.

Described from "Kirgiz deserts". Type in Berlin (B) ?

Rocky slopes in steppes and semideserts; plains and foothills.

IIA. Junggar: *Cis-Alt.* ("Bugotar brook on Sepske plateau, June 23, 1903"—Gr.-Grzh., l.c.); *Jung. Gobi* (nor.: left bank of Ch. Irtysh, 38 km from Shipati crossing on road to Koktogai, desert steppe on hummocky area, July 8, 1959—Yun. and I.-F. Yuan') *Zaisan* (Maikapchagai mountain, rocky slope, June 6, 1914—Schischk.); *Dzhark.* (Ili river west of Kul'dzha, May; Ili river south-west of Kul'dzha, May 29; Talki river gorge, July 18—1877, A. Reg.).

General distribution: Aralo–Casp., Fore Balkh.; Europe (south-east. European part of USSR), West. Sib. (south.; excluding Altay).

70. **A. glomeratum** Prokh. in Izv. Glavn. bot. sada SSSR, 29 (1930) 560; Vved. in Fl. SSSR, 4 (1935) 220; Fl. Kirgiz. 3 (1951) 84; Vved. in Opred. rast. Sr. Azii [Key to Plants of Mid. Asia], 2 (1971) 73. —Ic.; Prokh. l.c. tab. 2. Described from East. Pamir. Type in Leningrad. Plate IV, fig. 4.

Arid melkozem and rocky slopes, on talus, in montane steppes; middle and upper mountain belts.

IB. **Kashgar:** *Nor.* (Uchturfan, from Karol to Bedel pass, June 29, 1908—Divn.); *West.* (Kyude pass along Sinkiang-Tibet highway, No. 10079, July 27, 1959, Lee et al. (A.R. Lee (1959)).

IIA. **Junggar:** *Tarb.* (nor.-west of Khobsair mountain range, Chagan-Obo mountain, 2000 m, No. 10535, June 26, 1959—Lee and Chu); *Tien Shan* (along road to Karashar from Urumchi, valley terrace, 1900 m, No. 5937, July 21; along Urumchi-Karashar highway, on slope, 2300 m, Nos. 6106, 6107, July 22; same locality, 2800 m, No. 6120, July 22; nor. of Bagrashkul' lake, Lotogou region, arid slope, 2550 m, No. 06262, Aug. 1; Khotun-Sumbul district, Bain-Bulak vicinity, arid alluvial terrace, 2560 m, No. 6407, Aug. 6; 2 km south-west of Bain-Bulak, 2650 m, No. 6452, Aug. 10—1958, Lee and Chu); Khangi river valley 25 km north-west and beyond Balinte settlement, along road to Yuldus from Karashar, standing moraine, in gullies, Aug. 1; same locality, steppe on standing moraine, Aug. 1; B. Yuldus basin 4–5 km south-west of Bain-Bulak settlement, remnant hill along right bank of Khaidyk-Gol, along marbled limestone talus, Aug. 10; same locality, Khaidyk-Gol valley 3–4 km from Bain-Bulak settlement, steppe on pebble bed; Aug. 10; same locality, 10 km east of Bain-Bulak settlement along road to M. Yuldus, alpine steppe, Aug. 11—1958, Yun. and I.-F. Yuan').

IIIC. **Pamir** (Kok-Muinak settlement, clayey slope, July 27, 1909—Divn., typus!).
General distribution: Cent. Tien Shan.

71. **A. jacquemontii** Kunth, Enum. Pl. 4 (1843) 399; Henderson and Hume, Lahore to Jarkand (1873) 338; Hemsley, Fl. Tibet (1902) 199; Pampanini, Fl. Carac. (1930) 88; Wendelbo in Reching f. Fl. Iran. 76, 1, 4 (1971) 36. —Ic.: Wendelbo, l.c. tab. 4, fig. 45.

Described from India. Isotype in London (K).

Alpine belt.

IIIB. **Tibet:** *Chang Tang* ("Valle del Caracásh, ca. 4275 m"—Henderson and Hume, l.c.; "Lago Pàncong: fra il lago Pàn   cong e Mògaleb, 4358 m, Roero"—Pampanini, l.c.); *South.* (Ali: "Near Rakas Tal, 4500–5100 m, Strachey and Winterbottom"—Hemsley, l.c.).

General distribution: Fore Asia (Afghanistan, Pakistan), Himalayas (west.).

Note. The specimens cited from literature perhaps belong not to *A. jacquemontii* but *A. pamiricum* Wendelbo in Bot. Notiser 122, 1 (1969) 32. The latter species was described by Wendelbo from West. Pamir (USSR) and cited by him for this territory in place of *A. jacquemontii*. The latter was listed for West. Pamir by A.I. Vvedensky in Fl. SSSR, 4 (1935) 215, in Fl. Tadzh. 2 (1963) 322 and in Opred. rast. Sr. Azii [Key to Plants of Mid. Asia], 2 (1971) 71. In Fl. SSSR, A.I. Vvedensky remarked that he was assigning the Pamir plant to *A. jacquemontii* with some uncertainty because he had not studied its type. Wendelbo, who studied the isotype of *A. jacquemontii*, distinguishes it from *A. pamiricum* described by him in the following characters: in *A. jacquemontii*, the outer perianth lobes are broader than inner, filaments of stamens narrowly deltoid and anthers dark violet while, in *A. pamiricum*, the outer perianth lobes are narrower than inner, filaments of inner stamens broadly deltoid and anthers yellow.

72. **A. macrostemon** Bge. Enum. pl. China bor. (1832) 65; id. in Mém. Ac. Sci. St.-Pétersb. Sav. Étr. 2 (1835) 139; Rgl. Acta Horti petrop. 3, 2 (1875) 105; Forbes and Hemsley, Index Fl. Sin. 3 (1905) 123; Vved. in Fl. SSSR 4 (1935) 221; Kitag. Ling. Fl. Mansh. (1939) 131; Dashnyam in Bot. zh. 50, 11 (1965) 1639; Grub. in Novosti sist. vyssh. rast. 9 (1972) 276.—*A. uratense* Franch. Pl. David. 1 (1884) 304. —*A. macrostemon* var. *uratense* (Franch.) Airy-Shaw in Not. Roy. Bot. Gard. Edinb. 16 (1931) 136.

Described from North China. Type in Paris (P).

IA. **Mongolia:** *East. Mong.* (Ourato, July 1866—David; "Tsogtuzdelger mountain, meadow slopes of mud cones along gullies, 1965, Dashnyam"—Dashnyam, l.c.; Grub. l.c.).

General distribution: Far East, China (Dunbei, North), Japan.

73. **A. pallasii** Murr. Comment. Goetting. 6 (1775) 32, tab. 3; Rgl. in Acta Horti Petrop. 3, 2 (1875) 101 and 10, 1 (1887) 316, p.p.; Kryl. Fl. Zap. Sib. 3 (1929) 612; Vved. in Fl. SSSR, 4 (1935) 220; Fl. Kirgiz. 3 (1951) 84; Fl. Kazakhst. 2 (1958) 177; Grub. in Bot. mat. (Leningrad) 19 (1959) 534; Vved. in Opred. rast. Sr. Azii [Key to Plants of Mid. Asia], 2 (1971) 73. —*A. caricifolium* Kar. et Kir. in Bull. Soc. natur. Moscou, 14 (1841) 854. —*A. semiretschenkianum* Rgl. in Acta Horti Petrop. 5, 2 (1878) 630. —*A. albertii* Rgl. in Acta Horti Petrop, 5, 2 (1878) 632; Fl. Kazakhst. 2 (1958) 177.

Described from Siberia. Type in London (Linn.). Plate IV, fig. 1.

Sandy steppes, sand, melkozem-like and rubble slopes; plains and foothills to middle mountain belt.

IIA. **Junggar:** *Tien Shan* (Almaty gorge nor. of Kul'dzha, June 21; Khoyur-Sumun south of Kul'dzha, 1800–2100 m, May 26; Khanakhai, 1500–2100 m, June 16; Avraltau along Kash river, Aug. 5—1878, A. Reg.; mountains near Sairam lake, July 14, 1878—Fet.; Sygashu, May 4; Taldy river, 900–1050 m, June 15—1879, A. Reg.; Manas river near Chendokhodza river estuary, May 28, 1954—Mois.; nor. slope of East. Tien Shan, Urumchi town vicinity, south. extremity of Yaokhan' mountain range, on top of mound, June 3, 1957—Yun. et al.; San'daokhetsza region nor. of Sygashu, No. 31, June 4; near Ili river 25 km nor. of Syata, 1850 m, No. 1367a, Aug. 17—1957, Kuan; East. Tien Shan, nor. foothills of Bogdo-Ul hills 40–45 km east of Urumchi along road to Fukan, in gullies, April 26, 1959—Yun.; nor. slope of Merzbacher mountain range, upper Shigou river, uninundated erosion terrace, 2000 m, May 27, 1952—Mois.); *Jung. Gobi* (nor.: Qinhe-Ertai, No. 0728, July 29, 1956—Ching; 30 km south of Dinshan', on sand-dunes, Nos. 10304, 10305, May 28, 1959—Lee and Chu (A.R. Lee (1959)); west: along road to Paotai town from Savan town, No. 773, June 10, 1957—Kuan; south: Junggar plain adjoining Tien Shan debris cones, 24 km west of Shikhodze along road to Santokhodze, old wormwood fallow land, July 3, 1957—Yun. et al.); *Zaisan* (left bank of Ch. Irtysh, Dzhelkaidar area, sand, June 8; Besh-Kuduk wells, rocky-sandy steppe, June 18, 1914—Schischk.); *Dzhark.* (Ili river west of Kul'dzha, May; Kul'dzha, Apr. 20 and May 15; Pilyuchi near Kul'dzha, June 19, 1877,

A. Reg.; near Suidun on Ili river, 1878; Bayandai west of Kul'dzha, 600–900 m, May 6; Talki river near Suidun, May 7—1878, A. Reg.).

General distribution: Aralo-Casp., Fore Balkh., Jung.-Tarb., Nor. and Cent. Tien Shan; West. Sib (south.).

Note. A specimen grown from bulbs collected by A. Regel near Suidun on Ili river was described as A. albertii Rgl. l.c. This specimen has a subglobose inflorescence as in A. pallasii. In all other characters also, A. albertii is similar to this species.

74. A. sabulosum Stev. ex Bge. in Goebel. Reise Stepp. Russl. 2 (1838) 311; Regel in Izv. Obshch. lyubit. estestvozn. antrop. etnogr. 21, 2 (1876) 52; Rgl. in Acta Horti Petrop. 10, 1 (1887) 313; Vved in Fl. SSSR, 4 (1935) 228; Fl. Kazakhst. 2 (1958) 181; Vved. in Opred. rast. Sr. Azii [Key to Plants of Mid. Asia], 2 (1971) 76. —Ic.: Regel, l.c. (1876), tab. 9, fig. 1–3.

Described from Kazakhstan (Caspian region deserts). Type in Helsinki? Sandy deserts.

IIA. Junggar: Dzhark. (near Khorgos [May 13, 1878—A. Reg.]; Suidun, May 8; between Suidun and Ili river, June 2—1878, A. Reg.).

General distribution: Aralo-Casp., Fore Balkh.; Fore Asia (Iran), Mid. Asia (plain).

75. A. sairamense Rgl. in Acta Horti Petrop. 6, 2 (1880) 520 and 10, 1 (1887) 321. —A. kesselringii Rgl. l.c. 10, 1 (1887) 320. —A. schoenoprasoides auct. non Rgl.: Vved. in Fl. SSSR, 4 (1935) 224, p.p., quoad pl. chinens.; idem. in Opred. rast. Sr. Azii [Key to Plants of Mid. Asia], 2 (1971) 75, p.p. —Ic.: Rgl. l.c. 1887, tab. 1, fig. 3 (sub nom. A. kesselringii).

Described from Junggar (Tien Shan). Type in Leningrad. Plate IV, fig. 3.

Rocky and rubble slopes, talus and rocks; upper, rarely middle mountain belt.

IIA. Junggar: Jung. Alatau (20 km south of Ven'tsyuan' [Arasan], 2810 m, No. 1483, Aug. 14, 1957—Kuan); Tien Shan (Talki sources near Sairam lake, 3000 m, July 22, 1877—A. Reg., lectotypus!; south. slope of Bogdo mountains, July 24; Kokkamyr mountains, 2400–2700 m, July 24—1878, A. Reg.; from Bagaduslung to Bainamun, 1800 m, May 4; Borgaty pass, 2400–2700 m, June 7; Nilki brook of Kash river, 2100 m, June 8; Naryn-Gol, tributary of Tsagan-Usu, 1800–2400 m, June 10; Borborogusun river, 2700 m, June 15—1879, A. Reg.; Danu river valley, south of Danu town, on rocks, 2800 m, Nos. 348, 383, 2105, July 21; same locality, No. 530, July 22; Savan district, Datszymyao, subalpine meadow, No. 1722, July 22; mountains in Aksu town region, 3000 m, No. 3525, Aug. 14—1957, Kuan; Manas river-basin, upper course of Danu-Gol 1–2 km beyond opening to Danu pass, rocks and talus, July 21; upper course of Danu-Gol, in side opening to Danu pass, sheep's fescue-kobresia alpine steppe, July 23—1957, Yun. et al.).

General distribution: endemic.

Note. This species, closely related to A. schoenoprasoides Rgl., was formerly regarded as its synonym. The difference between the two species lies in the enlarged parts of filaments of inner stamens in A. sairamense cuneately narrowing upward and without teeth; in A. schoenoprasoides, they are enlarged upward and bear short distinct or faint teeth. This distinctive character of A. sairamense is persistent in all the specimens cited above.

76. **A. schoenoprasoides** Rgl. in Acta Horti Petrop. 5, 2 (1878) 630; Vved. in Fl. SSSR, 4 (1935) 224, p.p., excl. pl. chinens.; Fl. Kirgiz. 3 (1951) 86; Fl. Kazakhst. 2 (1958) 179; Fl. Tadzh. 2 (1963) 328; Ikonn. Opred. rast. Pamira [Key to Plants of Pamir] (1963) 87; Vved. in Opred. rast. Sr. Azii [Key to Plants of Mid. Asia], 2 (1971) 75, p.p., excl. syn. *A. sairamense* Rgl. and *A. kesselringii* Rgl.

Described from Kazakhstan (Tien Shan: Transili Alatau). Type in Leningrad.

Rocky and rubble slopes; upper part of middle and upper mountain belts.

IB. **Kashgar:** *West.:* (nor. slope of Kingtau mountain range, 3 km south-east of Kosh-Kulak settlement, along gullies, among junipers, June 10, 1959—Yun. and I.-F. Yuan').

IIA. **Junggar:** *Tien Shan* (Aryslyn, 2400–2700 m, Aug. 8, 1879—A. Reg.).

General distribution: Jung.-Tarb., Nor. and Cent. Tien Shan, East. Pam.; Fore Asia (nor.-east. Afghanistan), Mid. Asia (West. Tien Shan, Pam.-Alay).

77. **A. tanguticum** Rgl. in Acta Horti Petrop. 10, 1 (1887) 317, 287; Diels in Futterer, Durch Asien (1903) 7; Forbes and Hemsley, Index Fl. Sin. 3 (1905) 125. —**Ic.:** Rgl. l.c. tab. 2, fig. 1.

Described from Qinghai. Type in Leningrad.

Rocky steppe slopes, sandy banks and river falls; middle mountain belt.

IIIA. **Qinghai:** *Nanshan* (South Tetung mountain range, on sandy bank of river, Aug. 9, 1880—Przew., typus!; high foothills of Nanshan 60 km south-east of Chzhan'e town, 2200 m, July 12, 1958—Petr.; 33 km west of Xining town, montane wormwood-grass-forb steppe on rocky slopes, 2450 m, Aug. 5; 20 km west of Gunkhe town, steppe, 2980 m, Aug. 6—1959, Petr.; "Am Kuke-Nur selbst, August [undated], Futterer"—Diels, l.c.); *Amdo* (Mudzhik river valley in basin of upper course of Huang He river, swampy cliffs, June 29, 1880—Przew.).

General distribution: China (North-west: Gansu).

Note. The specimen collected in Mudzhik river valley differs from type in the presence of short teeth on enlarged filaments of inner stamens.

Section **Codonoprasum** (Reichb.) Endl.

78. **A. stamineum** Boiss. Diagn. pl. Or. Nov. sér. 2, 4 (1859) 119; Wendelbo in Rech. f. Fl. Iran. 76, 1, 4 (1971) 63; Vved. in Fl. SSSR, 4 (1935) 202. —*A. pseudoflavum* Vved. in Byull. Sredneaz. Univ. 19 (1934) 123; Vved. in Fl. SSSR, 4 (1935) 203. —*A. flavum* auct. non L.: Rgl. in Acta Horti Petrop. 3, 2 (1875) 187 and 10, 1 (1887) 350. —**Ic.:** Fl. SSSR, 4, Plate XII, fig. 1 (sub nom. *A. pseudoflavum*); Fl. Iran. 76, 1, 4, tab. 6, fig. 88, tab. 17, fig. 1.

Described from Europe (East. Mediterr.) and Fore Asia. Syntype in Geneva (G).

IIA. Junggar: *Dzhark.* (Kul'dzha, 1877—A. Reg.).
General distribution: Europe (East. Mediterr.), Caucasus (Transcaucasus), Fore Asia.

Note. The report of this Mediterranean-Fore Asian species in Kul'dzha is quite surprising and could be a labelling confusion.

There is no doubt about the identity of Kul'dzha plants with *A. stamineum*. They fully match with the specimens of this species available in the Herbarium of the Kamarov Botanical Institute as well as the description and drawings of Wendelbo (l.c.) who studied the type. *A. stamineum* represents a unique member of section Codonoprasum in Cent. Asia, its range encompassing Europe, the Mediterranean, Asia Minor, Fore Asia and partly Mid. Asia (Turkmenia).

Subgenus MOLLIUM (Koch) Wendelbo

Section Porphyroprason Ekberg

79. **A. oreophilum** C.A. Mey. in Verz. Pfl. Cauc. (1831) 37; Rgl. in Acta Horti Petrop. 3, 2 (1875) 210 and 10, 1 (1887) 355; Vved. in Fl. SSSR, 4 (1935) 255; Fl. Kirgiz. 3 (1951) 89; Fl. Kazakhst. 2 (1958) 186; Fl. Tadzh. 2 (1963) 338; Ekberg in Bot. Notiser, 122, 1 (1969) 66; Vved. in Opred. rast. Sr. Azii [Key to Plants of Mid. Asia], 2 (1971) 80. —*A. platystemon* Kar. et Kir. in Bull. Soc. natur. Mosc. 15, 3 (1842) 514. —*A. ostrowskianum* Rgl. in Acta Horti Petrop. 7 (1881) 545. —*A. oreophillum* β. *ostrowskianum* (Rgl.) Rgl. in Acta Horti Petrop. 10, 1 (1887) 356. —**Ic.:** Fl. Kazakhst. 2, Plate 14, fig. 3.

Described from Caucasus. Type in Leningrad.
Rocky and rubble slopes in upper mountain belt.

IIA. Junggar: *Jung. Alatau* ("Along Archaty river at its confluence with Borotala river, Sept. 11, 1878, Kushakevich"—Rgl. l.c. 1887).

General distribution: Jung.-Tarb., Nor. and Cent. Tien Shan; Fore Asia (nor.-east. Afghanistan and nor.-east. Pakistan), Caucasus, Mid. Asia (West. Tien Shan, Pam.-Alay).

Note. A specimen from Archaty river (Kushakevich's collections) was cited by E.L. Regel (l.c. 1887) as *A. oreophilum* β. *ostrowskianum* (Rgl.) Rgl. In Ekberg's article (l.c.) on *A. oreophilum*, a distribution map of this species was given (based on literature and herbarium data), according to which *A. oreophilum* also enters Junggar territory.

Subgenus MELANOCROMMYUM (Webb et Berth.) Wendelbo

Section Megaloprason Wendelbo

80. **A. decipiens** Fisch. ex Schult. et Schult. f. in Roem. et Schult. Syst. Veg. 7 (1830) 1117; Rgl. in Acta Horti Petrop. 3, 2 (1875) 245 and 10, 1

(1887) 359; Kryl. Fl. Zap. Sib. 3 (1929) 632; Vved. in Fl. SSSR, 4 (1935) 265, p.p., excl. syn. *A. roborowskianum* Rgl.; Fl. Kazakhst. 2 (1958) 189; Vved. in Opred. rast. Sr. Azii [Key to Plants of Mid. Asia], 2 (1971) 83. —*A. atropurpureum* auct. non Waldst. et Kit.: Rgl. l.c. 3, 2 (1875) 247, p.p. quoad pl. song.-kirghis. et excl. syn. *A. robustum*. —Ic.: Fl. Kazakhst. 3, Plate XIV, fig. 1.

Described from Europe (south. European part of USSR). Type in Berlin (B) ? Plate IV, fig. 7.

Slopes in lower hill belt.

IIA. **Junggar:** *Tien Shan* (Sary-Bulak near Kul'dzha, 1200 m, June 4; Kul'dzha, April 20; Almaty valley nor.-west of Kul'dzha, 900–1200 m, April 22; Almaty, Kokkamyr, April 28—1878, A. Reg.).

General distribution: Aralo-Casp., Fore Balkh., Jung.-Tarb.; Europe (south. European part of USSR), West. Sib. (south.).

81. **A. robustum** Kar. et Kir. in Bull. Soc. natur. Moscou, 14 (1841) 853; Vved. in Fl. SSSR, 4 (1935) 265; Fl. Kazakhst. 2 (1958) 189; Vved. in Opred. rast. Sr. Azii [Key to Plants of Mid. Asia], 2 (1971) 83. —*A. roborowskianum* Rgl. in Acta Horti Petrop. 10, 1 (1887) 359. —*A. atropurpureum* auct. non Waldst. et Kit.: Rgl. in Acta Horti Petrop. 3, 2 (1875) 247, p.p., quoad syn. *A. robustum*.

Described from East. Kazakhstan (Tarbagatai). Type in Leningrad. Plate IV, fig. 6.

Steppe mountain slopes; lower and middle mountain belts.

IIA. **Junggar:** *Tarb.* (nor.-west of Khobsair, Chaganobo mountain, on south. slope, 2000 m, No. 10562, June 22, 1959—Lee et al. (A.R. Lee (1959)); *Jung. Alatau* (Kerskenterek, May 24, 1878—A. Reg.); *Jung. Gobi* (east. Shankhausyan mountains, sandy places, May 24, 1879—Przew.); *Tien Shan* (Almaty valley nor.-west of Kul'dzha, 900–1200 m, April 22, 1878—A. Reg.).

General distribution: Fore Balkh. (Zaisan), Jung.-Tarb.

Note. This species is closely related to *A. decipiens*, from which it usually differs in thickened scape and atropurpureous colour of the perianth (a characteristic not usually detected in herbarium material), but mainly in the shape of notches between non-connate parts of filaments of stamens: in *A. decipiens*, the notches are broad, round or sublinear but in *A. robustum*, cuneate. With respect to shape of notches, *A. roborowskianum* Rgl. described from Junggar Gobi should be placed as a synonym of *A. robustum* and not *A. decipiens*, but the flowers of the type and solitary specimens of this species, judging from notes on the label, are pale lilac, as in *A. decipiens*.

Kanitz [in Széchenyi, Keletázsiai utjának, 2 (1981) 59; id. in Széchenyi, Wissensch. Ergebn. 2 (1898) 733] erroneously cites for Qinghai [Amdo ("Tsingtschou, Aug. 28, 1879, Széchenyi"] *A. fetissovii* Rgl., a species also belonging to section Megaloprason. *A. fetissovii* grows in Nor. and West. Tien Shan and its occurrence in Qinghai is therefore altogether improbable. Was *A. tanguticum* Rgl. mistaken for *A. fetissovii* ? (see section Scorodon, described above).

7. Lilium L.

Sp. pl. (1753) 302

1. Flowers erect, broadly funnel- or cup-shaped, with sublinear perianth lobes .... 2.
+   Flowers pendent, generally with perianth lobes strongly recurved .... 3.
2. Stem ribbed, thick, up to 1 m tall, with white flocculent pubescence. Flowers orange- or blood-red, rarely yellow, spotted, large; perianth lobes 4–8 cm long, clawed; pistil longer than stamens .... 2. **L. dauricum** Ker-Gawl.
+   Stem slender, smooth and glabrous, 20–60 cm tall. Flowers light-red or yellow, with dark speckles; perianth lobes 2.5–4 cm long, sessile; pistil much shorter than stamens .... 1. **L. buschianum** Lodd.
3. Leaves lanceolate, in whorls; some alternate. Flowers not fragrant; perianth lobes lilac, speckled .... 3. **L. martagon** L.
+   Leaves linear or linear-lanceolate, alternate. Flowers generally fragrant; perianth lobes red or white, without speckles .... 4.
4. Flowers bright red. Leaves linear .... 5.
+   Flowers white. Leaves linear-lanceolate .... 6. **L. tianschanicum** Ivanova.
5. Stem with glaucescent bloom, densely covered throughout length with short white papillae. Bulb yellowish-brown. Capsule globose .... 4. **L. potaninii** Vrishcz.
+   Stem usually green and glabrous or villose only at base. Bulb white. Capsule oblong-obovoid .... 5. **L. pumilum** Delile.

1. **L. buschianum** Lodd. Bot. Cab. 17 (1830) tab. 1628; Vrishcz, Lilii Dal'n. Vost. i Sib. [Lilies of the Far East and Siberia] (1972) 30; Grub. in Bot. zh. 56, 11 (1971) 1642. —*L. pulchellum* Fisch. in Index Sem. Horti Petrop. 6 (1839) 56; Danguy in Bull. Mus. nat. hist. natur. 20 (1914) 141; Kom. in Fl. SSSR, 4 (1935) 287. —*L. concolor* var. *buschianum* (Lodd.) Baker in J. Linn. Soc. (London), Bot. 14 (1874) 236; Forbes and Hemsley, Index Fl. Sin. 3 (1903) 129; Kitag. Lin. Fl. Mansh. (1939) 137. —Ic.: Lodd. Bot. Cab. 17, tab. 1628; Gartenfl. 9, tab. 284, fig. 2; Fl. SSSR, 4, Plate XVII, fig. 4.

Described from Dauria. Type—illustration in Lodd. Bot. Cab. 17 (1830) tab. 1628.

Dry valley and steppe meadows, grassy and sandy steppes.

IA. Mongolia: *East. Mong.* (Khailar town, wild rye steppe in sand, 1959—Ivan.). **General distribution:** East. Sib. (Daur.), Far East (south.), Nor. Mong. (Fore Hing.), China (Dunbei).

2. **L. dauricum** Ker-Gawl. in Curtis's Bot. Mag. 30 (1809) tab. 1210; Danguy in Bull. Mus. nat. hist. natur. 20 (1914) 141; Kom. in Fl. SSSR, 4

(1935) 284; Kitag. Lin. Fl. Mansh. (1939) 138; Grub. Konsp. fl. MNR (1955) 95. —*L. pensylvanicum* Ker-Gawl. in Curtis's Bot Mag. 22 (1804) tab. 872; Vrishcz, Lilii Dal'n. Vost. i Sib. [Lilies of the Far East and Siberia] (1972) 15. —**Ic.:** Curtis's Bot. Mag. 22, tab. 872; 30, tab. 1210; Gartenfl. 21, tab. 740.

Described from Dauria. Type—illustration in Curtis's Bot. Mag. 30 (1809) tab. 1210.

Forest and coastal meadows, forest fringes, among scrubs, in pine groves and larch forests.

IA. Mongolia: *East. Mong.* ("Environs de Kailar, alt. 760 m, monticules de sable, June 23, 1896, Chaff."—Danguy, l.c.; Khailar town, July 6, 1901—Lipsky; Khailar railway station, forest, June 10, 1902—Litw.; Khailar town vicinity, in scrubs at foot of sand-dunes, No. 783, June 20, 1951—S.H. Li et al.).

General distribution: East. Sib., Far East, Nor. Mong. (Hent., Mong.-Daur., Fore Hing.), China (Dunbei), Korean peninsula (nor.).

Note. There is no need following very formal considerations to reject the species name that has long been universally accepted in preference to the contentious *L. pensylvanicum* Ker-Gawl., which was acknowledged as erroneous by the author of the species himself and who proposed its replacement by the existing correct name *L. dauricum* Ker-Gawl.

3. **L. martagon** L. Sp. pl. (1753) 303; Danguy in Bull. Mus. nat. hist. natur. 17 (1911) 557 and 20 (1914) 141; Kom. in Fl. SSSR, 4 (1935) 288; Grub. Konsp. fl. MNR (1955) 95; Fl. Kazakhst. 2 (1958) 193; Pazij in Opred. rast. Sr. Azii [Key to Plants of Mid. Asia], 2 (1971) 89; Vrishcz, Lilii Dal'n. Vost. i Sib. [Lilies of the Far East and Siberia] (1972) 63. —*L. martagon* var. *pilosiusculum* Freyn in Oesterr. Bot. Zeitschr. 40 (1890) 224; Kryl. Fl. Zap. Sib. 3 (1929) 634. —*L. pilosiusculum* (Freyn) Miscz. in Tr. Bot. muz. Peterb. Akad. 8 (1911) 192; Vrishcz, Lilii Dal'n. Vost. i Sib. [Lilies of the Far East and Siberia] (1972) 60. —**Ic.:** Curtis's Bot. Mag. 23 (1805) tab. 893 and 39 (1814) tab. 1634; Reichb. Ic. Fl. Germ. 10, tab. 451, fig. 989.

Described from Europe and Siberia. Type in London (Linn.).

Larch and birch-larch forests, forest meadows and fringes, birch groves, meadow slopes of mountains.

IA. Mongolia: *Khobd.* (nor.-east. slope of Ulan-Daba pass, on dark side of gorge in forest, June 23, 1879—Pot.); *East. Mong.* ("Vallée du Keroulen, June 1896, Chaff."—Danguy, l.c.).

IIA. Junggar: *Cis-Alt.* (in Qinhe district [Chingil'], 1700–1800 m, No. 1221, Aug. 2, 1956—Ching); *Tarb.* (Dachen [Chuguchak], Kzyl-Tyunge, No. 1546, Aug. 12; north of Dachen, on slope, No. 2920, Aug. 13—1957, Kuan; nor.-west of Khobsair, Chagan-Obo hill, on exposed slope, 2000 m, No. 10552, June 22, 1959—Lee et al. (A.R. Lee (1959)).

General distribution: Jung.-Tarb.; Europe (Cent., South. and East.), Balk., West. Sib. (south.), East. Sib., Nor. Mong.

4. **L. potaninii** Vrishcz in Bot. zh. 53, 10 (1968) 1472; Vrishcz, Lilii Dal'n. Vost. i Sib. [Lilies of the Far East and Siberia] (1972) 75; Grub. in Bot. zh. 60, 7 (1975) 956. —*L. davidii* auct. non Duchartre: Krause in

Notizbl. Bot. Gart. Berlin, 9 (1926) 537. —*L. tenuifolium* auct. non Fisch.: Franch. Pl. David. 1 (1884) 307; Kanitz, A növenytani... (1891) 56; ej. Die Resultate... (1898) 734; Forbes and Hemsley, Index Fl. Sin. 3 (1903) 135, p.p.; Diels in Filchner, Wissensch. Ergebn. 10, 2 (1908) 248. —*L. tigrinum* auct. non Ker-Gawl.: Hao in Engler's Bot. Jahrb. 68 (1938) 588. —Ic.: Bot. zh. 53, 10, fig. 4.

Described from North-west China (Gansu). Type in Leningrad.

Scrubs, rocks and rocky steppe slopes, stable sand, arid sparse forests.

IA. Mongolia: *East. Mong.* (Muni-Ula, nor. slope, in birch grove, scattered, June 24, 1872—Przew.; 26 km south-west of Matad somon centre along road to Yugodzyr, rocky mountain slopes, July 26; Dariganga, Zodol-Khairkhan mountains, steppe along rocky volcano slopes, July 28, 1962—Yun. and Dashnyam; Dariganga, Dzun-Nart mound, south-west. slope, among scrubs on basalt outcrops, Aug. 10, 1970—Grub. et al.); *Alash. Gobi* (Alashan mountain range: cent. part of west. slope, among rocks, on melkozem, frequent, July 7, 1873—Przew.; Tszosto gorge, May 11; Khote-Gol gorge, south. slope, on sandy-humus soil, Nov. 12—1908, Czet.; near Baisa monastery, 30 km east of Bayan-Khoto town, juniper thickets, July 6; 50 km south-west of Inchuan' town, rocky slopes of mountain range, montane semidesert, July 10—1957, Petr.); *Ordos* (Ulan-Morin river valley, along nor. slopes of dunes among scrubs, Aug. 21, 1884—Pot.).

IIIA. Qinghai: *Nanshan* ("In ditione Tonkerr [Donkyr], July 16, 1879, Loczy"—Kanitz, l.c.; "Sining-fu, 1904, Filchner"—Diels, l.c., Krause, l.c.; "Sining-fu, Schangwuchuang, 2900 m, 1930"—Hao, l.c.; South Tetung mountain range, exposed slope, on humus, common, July 8, 1872; along Yusun-Khatym river, 2700–3050 m, on humus, July 23, 1880—Przew.; south. slope of Xining alps, Guiduisha, June 27, 1890—Gr.-Grzh.; Chortentan temple, 2100–2400 m, in spruce forests on humus, isolated plants, early Sept. 1901—Lad.; 70 km south-east of Chzhan''e town, Matisy temple, small meadows with scrub in hill valley, 2000 m, July 12, 1958—Petr.); *Amdo* (along Mudzhik river, 2700–2900 m, along cliff slope, scattered, June 30, 1880—Przew.).

General distribution: China (North and North-west).

5. **L. pumilum** Delile in Redouté, Liliac. 7 (1813) tab. 378; Edw. Bot. Reg. 2 (1816) tab. 132; Lodd. Bot. Cab. 4 (1819) tab. 388; Rgl. in Gartenfl. (1865) tab. 465; Vrishcz, Lilii Dal'n Vost. i Sib. [Lilies of the Far East and Siberia] (1972) 72. —*L. tenuifolium* Fisch. Catal. Hort. Gorenk. (1812) 8, nom. nud.; Fisch. ex Schranck, Pl. rar. horti acad. Monac. 2 (1819) tab. 91; Elwes, Monogr. Lilium, 7 (1880) tab. 42; Forbes and Hemsley, Index Fl. Sin. 3 (1903) 135, p.p.; Danguy in Bull. Mus. nat. hist. natur. 17 (1911) 557 and 20 (1914) 141; Kom. in Fl. SSSR, 4 (1935) 291; Kitag. Lin. Fl. Mansh. (1939) 139; Grub. Konsp. fl. MNR (1955) 95. —Ic.: Bot. Reg. 2, tab. 132; Bot. Mag. 59, tab. 3140; Gartenfl. 1865, tab. 465.

Described from a garden specimen in Paris. Type—illustration Redouté, Liliac. 7 (1813) tab. 378.

Steppes, rubble and rocky steppe slopes, rock crevices.

IA. Mongolia: *Cen. Khalkha* (Bichikte-Dulan massif [granite], clayey-rocky soil, Aug. 31, 1925—Gus.; upper Ubur-Dzhargalante river near Botoga mountains, rocky gully bed, Sept. 5, 1925—Krasch. and Zam.; Ikhe-Tukhum-Nur lake environs, Ulanche-Delger mountain, June 1926—Zam.; Sorgol-Khairkhan mountain, 180 km south-south-west of Ulan-Bator along old road to Dalan-Dzadagad, among crevices in pillow granites, July 15, 1943—Yun.; Bayan-Ula hill along old Ulan-Bator—Dalan-Dzadagad road, on granite rocks on south. slope, July 12, 1948—Grub.); *East. Mong.* (Khortitu, June 30, 1931—I. Kuznetsov; Inter fl. Chui gol et Iben gol, prope lacum [lat. 48°6′, long. 119°6′, alt. 670 m], June 17, 1873—Fritsche; "Environs de Kailar, alt. 760 m, June 24, 1876, Chaff."—Danguy, l.c.; Abder river, July 6; Bilyutai hill, July 16—1899, Pot. and Sold.; Khailar, July 6, 1901—Lipsky; Khailar railway station, forest, June 10, 1902—Litw.; Manchuria railway station vicinity, 1915—E. Nechaeva; same locality, Nanshan hill, 750–850 m, on mountain slope, in rock crevices, No. 869, June 24; south of Manchuria railway station, Erdaotszintszy village, 800 m, on hill slope, in rock crevices, No. 946, June 26—1951, Wang Chang et al.; Khailar town, genuine steppe on sand-dunes, 1959—Ivan.; left bank of Khalkhin-Gol around Derkhin-Tsagan-Obo mound, nor. slope of mound, 913 m, forb-grass steppe, Aug. 16, 1970—Grub. et al.).

General distribution: East. Sib. (south.), Far East (south), Nor. Mong., China (Dunbei).

6. L. tianschanicum Ivanova. —*L. ninae* Vrishcz in Bot. zh. 53, 10 (1968) 1468, quoad pl. tiansch.; Vrishcz, Lilii Dal'n. Vost. i Sib. [Lilies of the Far East and Siberia] (1972) 68, quoad pl. tiansch.

Bulbus albus subglobosus ca. 3 cm in diam., squamis carnosis numerosis. Caulis ca. 25 cm altus apice uniflorus, rectus, in parte superiore glaber, in parte inferiore sparse papillulatus; folia caulina patentia linearis sessilia acuta plana, media 8–10 cm longa, 2–5 mm lata. Flos nutans: perianthium album (in sicco !), tepalis oblongo-lanceolatis, 4.5 cm longis, 12–14 mm latis, revolutis, apice callosis intus sparse papillulatis in sulco nectarifero dense papillatis; stamina perianthio subaequilonga, antherae flava.

Typus: declivitas borealis jugi Tian-Schan inter oppido Fukan et Monte Bogdo-Ola in steppa argilloso-glareosa, No. 1316, Aug. 2–3, 1908, Merzbacher. In Herb. Inst. Bot. Ac. Sci. URSS (Leningrad) conservatur.

IIA. **Junggar:** *Tien Shan* (Turkestania chinensis. Am Wege von Fucan zur Bogdo-Ola, in der Lehm- und Kiessteppe, No. 1316, Aug. 2–3, 1908—Merzb., typus!).

General distribution: endemic.

Note. The fairly distinct morphological differences (perianth white, without speckles, subglabrous stem, etc.) and the vast discontinuity of distribution ranges do not support clubbing this species with Sichuan *L. ninae* Vrishcz.

8. **Rhinopetalum** Fisch. ex Alexand.
in Edinb. New Philosoph. J. 8 (1930) 19.

1. Rh. karelinii Fisch. l.c.; Rgl. in Acta Horti Petrop. 6, 2 (1880) 514; Losinsk. in Fl. SSSR, 4 (1935) 297; Fl. Kazakhst. 2 (1958) 195; Fl. Tadzh. 2 (1963) 242; Pazij in Opred. rast. Sr. Azii [Key to Plants of Mid. Asia], 2 (1971) 90. —*Fritillaria karelinii* Rgl. ex Baker in J. Linn. Soc. (London) Bot.

14 (1875) 268; Rgl. in Acta Horti Petrop. 8 (1884) 652. —Ic.: Gartenfl. 1877, tab. 796; Fl. Kazakhst. 2, Plate 15, fig. 3 (sub *Rh. stenanthero*).

Described from West. Kazakhstan (Aralo-Casp.). Type in Leningrad. Sandy and clayey deserts.

IIA. Junggar: *Dzhark.* (sand near Suidun, 600 m, March 6–7; Suidun, May 8; Khorgos, June 9; Kul'dzha, April; Togustorau west of Kul'dzha, April; Pilyuchi, April; Bayandai plant, 600 m, May 5; Bayandai around Kul'dzha, 600–1200 m, May 6— 1879, A. Reg.).

General distribution: Aralo-Casp., Fore Balkh.; Mid. Asia.

9. Fritillaria L.

Sp. pl. (1753) 303.

1. Leaves opposite or in whorls of 3–4, only uppermost leaves alternate, linear-lanceolate or linear, often with curved tip. Flowers single, rarely in pairs .... 2.

+ All leaves alternate, broadly lanceolate, with erect tip. Flowers 2– 5 in terminal sparse raceme, rarely single, apple-green, with indistinct brownish chequered pattern inside ....
.... 1. **F. pallidiflora** Schrenk.

2. Upper leaves and bracts cirrhous at tip, lower ones obtuse. Flowers light coloured, yellow or albescent-green. Capsule acutely-angled .... 3.

+ Leaves opposite or in whorls, linear-lanceolate, with acute erect tip. Flowers violet-purple with atropurpureous spots. Capsule obtusely-angled .... 3. **F. soulie** Franch.

3. Lower leaves opposite, upper leaves alternate, linear. Flowers yellow with orange spots outside and dispersed black dots inside
.... 2. **F. przewalskii** Maxim.

+ Lower leaves opposite, middle leaves in whorls and upper opposite and alternate, linear-lanceolate. Flowers albescent-green, brownish-purple with white spots inside ..... 4. **F. walujewii** Rgl.

1. **F. pallidiflora** Schrenk in Fisch. et Mey. Enum. pl. nov. 1 (1841) 5 and 2 (1842) 5; Rgl. in Acta Horti Petrop. 6, 2 (1880) 513; Losinsk. in Fl. SSSR, 4 (1935) 304; Fl. Kazakhst. 2 (1958) 196; Pazij in Opred. rast. Sr. Azii [Key to Plants of Mid. Asia], 2 (1971) 93. —Ic.: Gartenfl. 1857, tab. 209.

Described from East. Kazakhstan (Jung. Alatau). Type in Leningrad.

Meadow slopes of mountains, rock terraces and ledges, upper mountain belt.

IA. Junggar: *Jung. Alatau* (30 km west of Arasan, 2640 m, on slope, No. 2028, Aug. 25, 1957—Kuan); *Tien Shan* (Talki brook, 900–1500 m, April 1877; Talki gorge, 1800–2400 m, Aug. 16, 1877; Sary-Bulak near Kul'dzha, 1200–1800 m, April 22 and 23; Kyzemchek mountains near Sairam lake, 3050 m, April 24; Almaty gorge near

Kul'dzha, April 26; Khorgos midcourse, 900–1500 m, May 15—1878; Borborogusun, 1500 m, April 27; Nilki brook, Kash river tributary, 2100 m, June 8—1879, A. Reg.; bei Manass, No. 1257, July 5–17, 1908—Merzb.).

General distribution: Jung.-Tarb., Nor. Tien Shan ?

2. F. przewalskii Maxim. in Trautvetter, Regel, Maximowicz et Winkler, Decas pl. nov. (1882) 9; Batal. in Acta Horti Petrop. 13 (1893) 105; Forbes and Hemsley, Index Fl. Sin. 3 (1903) 137.

Described from Qinghai. Type in Leningrad.

Meadows and along river banks in upper mountain belt.

IIIA. **Qinghai**: *Nanshan* (Xining hills, Myn'dan'sha river, No. 133, May 27 and No. 166, June 14, 1890—Gr.-Grzh.); *Amdo* (In montib. Mudshik, 3000 m, ad pedem praerupti argillosi secus rivum, solo humido, rara, No. 309, June 5 [17], 1880—Przew., typus!).

General distribution: endemic.

3. F. soulie Franch. in J. Bot. (Paris) 12 (1898) 221; Forbes and Hemsley, Index Fl. Sin. 3 (1903) 137. —*F. roylei* auct. non Hook.: Rehd. in J. Arn. Arb. 14 (1933) 6.

Described from South-west China. Type in Paris (P).

Alpine meadows.

IIIA. **Qinghai**: *Amdo* (Radja and Yellow river gorges: alpine region between Radja and Jupar range: alpine meadows of Wajola, alt. 14,000 ft [4250 m], No. 14160, June 1926—J. Rock, sub. *F. roylei*).

General distribution: China (South-west: Sichuan).

4. F. walujewii Rgl. in Gartenfl. 28 (1879) 353; id. in Acta Horti Petrop. 6, 2 (1880) 298 and 514; Losinsk. in Fl. SSSR, 4 (1935) 313; Fl. Kirgiz. 3 (1951) 97; Grub. in Bot. mat. (Leningrad) 19 (1959) 536; Pazij in Opred. rast. Sr. Azii [Key to Plants of Mid. Asia], 2 (1971) 93. —**Ic.**: Gartenfl. 28, tab. 993.

Described from Mid. Asia. Type in Leningrad.

Rocky and rubble slopes, among rocks in middle and upper mountain belts.

IA. **Junggar**: *Jung. Alatau* (Toli district, Albakzin, on slope, No. 2613, Aug. 7; 10 km south of Arasan, 1400 m, on slope, No. 1186, Aug. 23—1957, Kuan); *Tien Shan* (Tsanma river upper course, 2100 m, on rocks, rare, June 19, 1877—Przew.; Tekes valley, 1500 m, June 23; along Sharysa river, 1800 m, June 1878—A. Reg.; Iren-Khabirga mountain range: Taldy gorge inlet, 900–1200 m, May 14; Taldy river lower course, 1200 m, May 15; Taldy gorge [Turgen or Epte] 1500–2400 m, May 15; upper Taldy, 2100 m, May 15; Taldy lateral gorge, 2100–2700 m, May 15; Taldy river upper course, 2400–2700 m, May 17; Taldy river source, 3050 m, May 21, 1879; Aryslyn gorge, 2400–2700 m, July 10, 1879—A. Reg.; Santai, upper Shigou [Tshkasu] river, unflooded erosion terrace, 2000 m, May 27, 1952—Mois.; Chendokhoze river [Manas tributary], 2100–2300 m, May 27, 1954—Mois.; Urumchi town, Nanshan, 1100 m, No. 514, July 20, 1956—Ching; Kizylzangi, south of Shichan, No. 764, July 23; Bogdo-Ula, 15 km south of Tyan'chi lake, 1450 m, on nor.-west. slope, No. 1900, Sept. 18—1957, Kuan).

88

General distribution: Jung.-Tarb., Nor. and Cent. Tien Shan; Mid. Asia (Pam.-Alay, West. Tien Shan).

10. Tulipa L.
Sp. pl. (1753) 305; Gen. pl. (1754) 145.

1. Stigma sessile, style absent; filaments of stamens glabrous .... 2.
+ Stigma on well-developed style, as long as ovary; filaments of stamens glabrous or pilose .... 7.
2. Stem in upper part and pedicels pubescent. Flowers pure yellow
 .... 3.
+ Stem in upper part and pedicels glabrous .... 4.
3. Bulb tunic pilose throughout inner surface. Leaves 3 (4), from lanceolate to elliptical, chondroid and crispate along margin .....
 .... 1. **T. altaica** Pall.
+ Bulb tunic with appressed hairs inside only at very tip and base. Leaves (3) 4, linear to linear-lanceolate, with uniform margin .....
 .... 6. **T. iliensis** Rgl.
4. Perianth lobes yellow, without spots at base; filaments of stamens gradually narrowing from base. Leaves (2) 3. Flowers solitary ....
 .... 5.
+ Perianth lobes yellow, outer ones with violet spot or tinge outside at base; filaments of stamens narrow sharply into fine tip. Leaves 3–6, linear. Flowers sometimes in pairs .... 6.
5. Leaves nearly closely whorled, broadly linear, grooved, with uniform margin .... 9. **T. triphylla** Rgl.
+ Leaves distant, linear-lanceolate, with crispate margin ....
 .... **T. kolpakowskiana** Rgl.
6. Perianth lobes similar in shape, subacute or obtuse. Leaves 3–4, sometimes 5–6, strongly proximate, strap-shaped, crispate ....
 .... 7. **T. tetraphylla** Rgl.
+ Inner perianth lobes with rounded tip and with linear cusp. Leaves 3, distant faintly undulate along margin ....
 .... 2. **T. aristata** Rgl.
7. Filaments of stamens glabrous, anthers without appendage; flowers solitary. Bulb tunic thin, papery. Leaves strongly proximate .... 8.
+ Filaments of stamens pilose or with hairy ring at base, anthers with cusp. Bulb tunic coriaceous or papery. Leaves 2, distant, recurved .... 10.
8. Leaves 3 (rarely 2), falcate and proximate in highly elongated whorl, with cap at tip and finely dentate-villous along margin. Bulb tunic appressed-bristly-pilose at tip, strongly extended. Style short .... 8. **T. tianschanica** Rgl.

+   Leaves 2, glabrous and smooth. Style long .... 9.
9. Bulb tunic glabrous inside. Leaves opposite. Flower slightly open, funnel-shaped, greenish-bluish outside, white inside and along lobe margin; anthers about 2.5 mm long ....
.... 5. **T. heterophylla** (Rgl.) Baker.
+   Bulb tunic with appressed hairs inside tip. Leaves closer. Flowers wide open, cup-shaped, green outside, yellow inside; anthers 3–5 mm long .... 10. **T. uniflora** (L.) Bess.
10. Bulb tunic coriaceous, woolly inside tip. Flowers 2–3 and up to 6, white with yellow base; anthers 4–6 mm long ....
.... 3. **T. buhseana** Boiss.
+   Bulb tunic papery, glabrous inside or with few appressed hairs at very tip. Flower solitary, pale yellow; anthers 3–5 mm long ....
....4. **T. dasystemon** Rgl.

1. **T. altaica** Pall. ex Spreng. Syst. 2 (1825) 63; Rgl. in Acta Horti Petrop. 6, 2 (1880) 506; Kryl. Fl. Zap. Sib. 3 (1929) 638; Vved. in Fl. SSSR, 4 (1935) 346; Fl. Kazakhst. 2 (1958) 206; Vved. in Opred. rast. Sr. Azii [Key to Plants of Mid. Asia], 2 (1971) 104. —Ic.: Ledeb. Ic. pl. fl. ross. 2 (1830) tab. 134; Gartenfl. 27 (1878) tab. 942.

Described from Altay. Type in Leningrad.

Rubble and rocky steppe slopes of mountains and knolls.

IIA. **Junggar:** *Tien Shan* (Kul'dzha, 1876—Golike; Karkarausu, south-east of Shikho, 600 m, March 29, 1878—A. Reg.).

General distribution: Fore Balkh. (Zaisan), Jung.-Tarb.; West. Sib. (Altay: southwest.).

2. **T. aristata** Rgl. in Acta Horti Petrop. 6, 2 (1880) 506.

Described from Junggar. Type in Leningrad.

Steppe slopes.

IIA. **Junggar:** *Tien Shan* (Karkaraussu südostl. von Schicho, 2000 ft [600 m], March 29, 1878—A. Reg., typus!).

Note. Known only from a single authentic specimen; perhaps, only a random variety of *T. kolpakowskiana* Rgl. as admitted by E. Regel himself.

3. **T. buhseana** Boiss. Diagn. pl. or., ser. 2, 4 (1859) 98; Vved. in Fl. SSSR, 4 (1935) 356; Fl. Kazakhst. 2 (1958) 213; Vved. in Opred. rast. Sr. Azii [Key to Plants of Mid. Asia], 2 (1971) 107. —*T. turkestanica* auct. non Rgl. 1875; Rgl. in Acta Horti Petrop. 6, 2 (1880) 502, p.p. —Ic.: Fl. Kazakhst. 2, Plate XVII, fig. 3.

Described from Fore Asia (Iran). Type in Geneva (G).

Sandy and clayey soils in deserts.

IIA. **Junggar:** *Jung. Gobi* (nor.: along Urungu river, occasional, April 23 [May 5] 1879—Przew., cum *T. uniflora*) *Dzhark.* (Khorgos, April 22, 1877; Suidun, May 8, 1878—A. Reg.).

General distribution: Aralo-Casp., Fore Balkh.; Mid. Asia (plain, Kopet-Dag).

4. **T. dasystemon** Rgl. in Acta Horti Petrop. 6, 2 (1880) 507; Vved. in Fl. SSSR, 4 (1935) 361; Fl. Kirgiz. (1951) 115; Fl. Kazakhst. 2 (1958) 215; Fl. Tadzh. 2 (1963) 269; Vved. in Opred. rast. Sr. Azii [Key to Plants of Mid. Asia], 2 (1971) 108. —*Orithyia dasystemon* Rgl. in Acta Horti Petrop. 5 (1877) 261. —Ic.: Fl. Kazakhst. 2, Plate XVIII, fig. 2.

Described from East. Kazakhstan (Nor. Tien Shan). Type in Leningrad.

Steppe and meadow slopes from middle to upper (alpine) mountain belts.

IIA. Junggar: *Tien Shan* (Sairam-Kyzemchek, 3050 m, July 1877; Sarybulak nor.-west of Kul'dzha, 1200-1800 m, April 23; Talki brook, 900-1200 m, May 9; Dzhagastai river, 1300-2100 m, June 16—1878; upland at sources of Dzhirgalan and Pilyuchi, 1800 m, April 24, 1879—A. Reg.).

General distribution: Nor. Tien Shan; Mid. Asia (West. Tien Shan, Pam.-Alay).

5. **T. heterophylla** (Rgl.) Baker in J. Linn. Soc. (London) Bot. 14 (1874) 295; Rgl. in Acta Horti Petrop. 6, 2 (1880) 507; Vved. in Fl. SSSR, 4 (1935) 364; Fl Kirgiz. 3 (1951) 116; Fl. Kazakhst. 2 (1958) 216; Vved. in Opred. rast. Sr. Azii [Key to Plants of Mid. Asia], 2 (1971) 109. —*Orithyia heterophylla* Rgl. in Bull. Soc. natur. Moscou, 41, 1 (1868) 440. —Ic.: Izv. Obshch. lyubit. est., antrop. i etnogr. 21, 2 (1876) [Regel, Fl. Turk.] Plate 21, figs. 11–12.

Described from East. Kazakhstan (Nor. Tien Shan). Type in Leningrad.

Rubble and rocky slopes in steppes, in forests and alpine meadows.

IIA. Junggar: *Jung. Alatau* (20 km south of Ven'tsyuan' [Arasan], 2810 m, No. 1478, Aug. 14, 1957—Kuan); *Tien Shan* (Koshety-Daban, alps above forest line, June 11, 1877—Pot.; Muzart. 2700 m, Aug.—1877; south-east of Ketmen' pass, 2400-2700 m, June 19; Dzhagastai hill, 2400-2700 m, June 20—1878; Taldy river gorge opening, 900-1200 m, May 14; upper Taldy, 2100-2700 m, May 15; Taldy river upper course, 2400-2700 m, May 17; upper Taldy river, 2700-3050 m, May 19; Taldy sources, 3050 m, May [20], 1879; Borborogusun, 1800 m, April 2; same locality, 2700 m, June 15; nor. slope of Iren-Khabirga, Dumbedan—Kum-Daban, 2400-2700 m, May 28; Kumbel', 2700-3050 m, May 30; Bagaduslun, east. tributary of Dzhina, 2100-2700 m, June 4; Aryslyn, 2700-3050 m, July 16—1879, A. Reg.; 10 km nor. of Chzhaos [Kalmakkure], 2900 m, on mountain top, No. 3313, Aug. 15, 1957—Kuan; 1–2 km above Torugart settlement toward border, alpine meadow, June 20, 1959—Yun., Murzaev and I.-F. Yuan').

General distribution: Nor. and Cent. Tien Shan.

6. **T. iliensis** Rgl. in Gartenfl. 28 (1879) 162 and 227; id. in Acta Horti Petrop. 6, 2 (1880) 506; Vved. in Fl. SSSR, 4 (1935) 347; Fl. Kazakhst. 2 (1958) 206; Grub. in Bot. mat. (Leningrad), 19 (1959) 536; Vved. in Opred. rast. Sr. Azii [Key to Plants of Mid. Asia], 2 (1971) 104. —Ic.: Gartenfl. 28, tab. 975, figs. c–d, tab. 982, figs. 4–6.

Described from East. Kazakhstan (Tien Shan). Type in Leningrad.

Steppe and desert loessial and clayey slopes, river terraces, in wormwood groves and among steppe scrub.

IB. **Kashgar:** *Nor.* (22 km south of Baichen, 900 m, No. 2066, Sept. 22, 1957—Shen'-Tyan').

IIA. **Junggar:** *Tien Shan* (Sarybulak river nor.-west of Kul'dzha, 1200–1800 m, April [22–24]; Almaty valley near Kul'dzha, 900–1800 m, April 25; Almaty gorge, 750–900 m, April; Talki brook, 900–1500 m, May 9—1878; Pilyuchi gorge, 900–1200 m, April 22, 1879—A. Reg.; Kul'dzha, 1906—Muromskii; Kungess Tal, No. 1135, May 1–5, 1908—Merzb.; Santai, upper Shigou river, uninundated terrace, 2000 m, May 27, 1952—Mois.; Chendokhodza, 2100–2300 m, Palaeozoic rocks, May 27, 1954—Mois.; 10–15 km nor.-east of Urumchi along road to Fukan, wormwood groves along loessial ridges, April 26; same locality, 40–45 km from Urumchi, wormwood groves on eroded ridges, along gullies, April 26; 15–20 km from Fukan along lake road [Tyan'chi], steppe belt, in spirea thickets, fairly abundant, April 26—1959, Yun.; Fukan-Tyan'chi lake, 920 m, on nor. slopes of ridges, No. 1885, Sept. 18, 1959—A.R. Lee (1959); *Jung. Gobi.* (south.: in Kuitun-Yanzykhai interfluvial region, about 600 m, semidesert plain, loessial loam, April 26, 1954—Mois.; south of Savan 30 km away, along roadside, No. 733, June 10; along highway in Kuitun region, Gobi, No. 1090, June 27—1957, Kuan; 5 km nor. of Kuitun, 400 m, No. 10209, April 22, 1959—A.R. Lee (1959)); *Dzhark.* (Bayandai, west of Kul'dzha, 600–1200 m, May 6; between Suidun and Ili river, May 7—1878, A. Reg.).

General distribution: Nor. Tien Shan (Ketmen'tau foothills).

Note. A single report of this tulip in Kashgar (barring a labelling error, which is very unlikely) remains somewhat an enigma. No tulip species has been reported from Kashgar to date.

T. kolpakowskiana Rgl. in Acta Horti Petrop. 5, 1 (1877) 266 and 6, 2 (1880) 504, p.p.; Vved. in Fl. SSSR, 4 (1935) 345; Fl. Kirgiz. 3 (1951) 104, excl. syn.; Fl. Kazakhst. 2 (1958) 205; Vved. in Opred. rast. Sr. Azii [Key to Plants of Mid. Asia], 2 (1971) 103. —Ic.: Gartenfl. 27, tab. 951; Fl. Kazakhst. 2, Plate XVI, fig. 3.

Described from East. Kazakhstan (Nor. Tien Shan). Type in Leningrad. Loessial slopes of steppe and desert foothills.

Reliable specimens are not available from Junggar territory but the species is found in the border regions of the USSR. See also note under *T. aristata* Rgl. and *T. triphylla* Rgl.

General distribution: Jung.-Tarb. (south. Jung. Alatau), Nor. Tien Shan.

7. T. tetraphylla Rgl. in Acta Horti Petrop. 3 (1875) 296 and 6, 2 (1880) 503; Vved. in Fl. SSSR, 4 (1935) 350; Fl. Kirgiz. 3 (1951) 108; Fl. Kazakhst. 2 (1958) 208; Vved. in Opred. rast. Sr. Azii [Key to Plants of Mid. Asia], 2 (1971) 105. —*T. kesselringii* Rgl. in Acta Horti Petrop. 5, 2 (1878) 637. —Ic.: Gartenfl. 28, tab. 964 (sub *T. kesselringii*).

Described from Kirghizstan (Cent. Tien Shan). Type in Leningrad. Rocky slopes of hills.

IIA. **Junggar:** *Tien Shan* (Declivitas australis jugi montium Tianschan: Tilbitschuk Tal, No. 459, May 10, 1903—Merzb.); *Jung. Gobi* (south.: 5 km nor. of Kuitun, 400 m, No. 10210, April 22, 1959—Lee and Chu (A.R. Lee (1959)).

General distribution: Nor. and Cent. Tien Shan.

8. **T. tianschanica** Rgl. in Acta Horti Petrop. 6, 2 (1880) 508 and 8 (1884) 652; Vved. in Fl. SSSR, 4 (1935) 349; Fl. Kirgiz. 3 (1951) 108; Fl. Kazakhst. 2 (1958) 208; Vved. in Opred. rast. Sr. Azii [Key to Plants of Mid. Asia], 2 (1971) 105. —Ic.: Acta Horti Petrop. 8, tab. 5, fig. 3, f-k; Fl. Kazakhst. 2, Plate XVI, fig. 1.

Described from Junggar (Tien Shan). Type in Leningrad.

Rocky slopes in upper mountain belt.

IIA. **Junggar:** *Tien Shan* (M. Yuldus upland, 2300–2700 m, May 21 [June 2], 1877—Przew.; fl. Agias, 2100–2400 m, June 26, 1878—A. Reg., typus!; upper Taldy, 2400 m, May 25; south. source of Taldy, 2400 m, May 25; middle Taldy, 2100 m, May 26; Dumbedan—Kum-Daban, nor. slope of Iren-Khabirga, 2400–2700 m, May 28; Kum-Daban, 2700 m, May 29; Aryslyn, July 15—1879, A. Reg.; hohes Plateau eines Berges im oberen Koksu Gebiet, ca. 3400 m, No. 903, July 8–10, 1907; Kungess Tal, No. 1111, Sept. 1–5, 1908—Merzb.).

General distribution: Nor. and Cent. Tien Shan.

9. **T. triphylla** Rgl. in Acta Horti Petrop. 5, 2 (1878) 636 and 6, 2 (1880) 505, p.p., quoad pl. iliens.; id. in Gartenfl. (1878) 193, tab. 972.

Described from plants grown from bulbs collected in Junggar. Type in Leningrad.

On sand (?).

IIA. **Junggar:** *Dzhark.* ("In collibus arenosis inter Kuldscha et lacum Sairam", July 1877; "In valle fl. Ili prope Bajandai", July 11, 1877—A. Reg.).

General distribution: endemic (?).

Note. Very closely related to *T. kolpakowskiana* Rgl. and perhaps only a variety of it.

10. **T. uniflora** (L.) Bess. ex Baker in J. Linn. Soc. (London) Bot. 14 (1874) 295; Rgl. in Acta Horti Petrop. 6, 2 (1880) 506 cum auct. D. Don, p.p.; Vved. in Fl. SSSR, 4 (1935) 363; Grub. Konsp. fl. MNR (1955) 95; Fl. Kazakhst. 2 (1958) 216; Vved. in Opred. rast. Sr. Azii [Key to Plants of Mid. Asia], 2 (1971) 109.—*Ornithogalum uniflorum* L. Mant. (1767) 62. —*Orithyia uniflora* D. Don in Brit. Flow. Gard. 4 (1838) 336; Kryl. Fl. Zap. Sib. 3 (1929) 640.

Described from Altay. Type in London (Linn.).

Rubble and rocky slopes of mountains and knolls, sandy places, from desert foothills to upper mountain belt.

IA. **Mongolia:** *Mong. Alt.* (Dzuilin-Gola valley, pass 2925 m south of Sutai-Ula, wormwood-sheep's fescue-wheat grass montane steppe, June 23, 1971—Grub. et al.); *Depr. Lakes* (Khungui II valley 10 km nor.-east from Erdene-Khairkhan somon centre, on fringe of sand-dunes under Tsakhir-Ula hill, on sand with pea shrub cover, Aug. 20, 1972—Grub. et al.).

IIA. **Junggar:** *Cis-Alt.* (Bugotar brook, high plateau [Sepske] between Burchum and Kran [meadow] June 10 [22]; Kran river valley [in midcourse, meadow] June 23 [July 5]—1903, Gr.-Grzh.); *Jung. Alatau* (10 km south of Ven'tsyuan, in mountain, 1400 m, on exposed slope, No. 1185, Aug. 23, 1957—Kuan); *Tien Shan* (nor. slope of Bogdo Ula along road to lake [Tyan'chi] from Fukan, lower part of forest belt, on south. rocky steppe slope, April 26, 1959—Yun. and I.-F. Yuan'); *Jung. Gobi* (along

Urunga river, May 5; plain between Altay and Barkul, Kuku-Syrkhe mountains, May 21; Kholyt mountains, clayey-sandy soil, May 31—1879, Przew.; left bank of Ch. Irtysh 38 km east of Shipati crossing on road to Koktogai, desert steppe along hummocky area, July 8, 1959—Yun. and I.-F. Yuan'); *Zaisan* (?).

General distribution: Tarb., Fore Balkh. (Zaisan); West. Sib. (Altay), East. Sib. (Ang.-Sayan. south., Daur. south.), Nor. Mong. (Hangay).

11. Erythronium L.
Sp. pl. (1753) 305.

1. **E. sibiricum** (Fisch. et Mey.) Kryl. Fl. Zap. Sib. 3 (1929) 641; Krasch. in Fl. SSSR, 4 (1935) 365; Fl. Kazakhst. 2 (1958) 217. —*E. dens canis* var. *sibiricum* Fisch. et Mey. in Index sem. Horti Petrop. 7 (1841) and in Linnaea, 15 (1841) 111. —**Ic.:** Fl. SSSR, 4, Plate XX, fig. 1; Fl. Kazakhst. 2, Plate XIX, fig. 4.

Described from Altay. Type in Leningrad.

Forest and alpine meadows, thawed snow patches.

IIA. Junggar: *Cis-Alt.* (south. slope of Altay, Kran river [in midcourse, meadow], June 23 [July 5], 1903—Gr.-Grzh.).

General distribution: West. Sib. (Altay), East. Sib. (Sayans).

12. Lloydia Reichb.
Fl. Germ. Excurs. (1830) 102, nom. cons.

1. Radical leaves 1–2, filiform-linear, up to 1.5 mm broad. Flowers usually solitary, rarely 2–3; filaments of stamens glabrous, anthers oval .... 1. **L. serotina** (L.) Reichb.

+   Radical leaves 3–5, linear, herbaceous, up to 4 mm broad. Flowers generally 3–8, rarely solitary; filaments of stamens villous, anthers oblong ....2. **L. tibetica** Baker.

1. **L. serotina** (L.) Reichb. Fl. Germ. Excurs. (1830); 102; Rgl. in Acta Horti Petrop. 6, 2 (1880) 513; Danguy in Bull. Mus. nat. hist. natur. 14 (1908) 132; Ostenf. and Pauls. in Hedin, South. Tibet, 3, 3 (1922) 89; Kryl. Fl. Zap. Sib. 3 (1929) 642; Kom. in Fl. SSSR, 4 (1935) 369; Kitag. Lin. Fl. Mansh. (1939) 139; Fl. Kirgiz. 3 (1951) 116; Grub. Konsp. fl. MNR (1955) 95; Fl. Kazakhst. 2 (1958) 218; Fl. Tadzh. 2 (1963) 270; Ikonnik. Opred. rast. Pamira [Key to Plants of Pamir] (1963) 88; Pazij in Opred. rast. Sr. Azii [Key to Plants of Mid. Asia], 2 (1971) 109. —*L. alpina* Salisb. in Trans. Hort. Soc. i (1812) 328; Forbes and Hemsley, Index Fl. Sin. 3 (1903) 139, p.p. —*Bulbocodium serotinum* L. Sp. pl. (1753) 294. —**Ic.:** Fl. Kazakhst. 2. Plate XIX, fig. 3; Tadzh. 2, Plate 52, fig. 2.

Described from Europe (Swiss Alps). Type in London (Linn.).

Crevices and rock ledges, humid marshy meadows, rocky and rubble placers, moraine and talus in alpine belt.

IA. Mongolia: *Khobd.* (Kharkhira and Turgen' mountain ranges); **Mong. Alt.,** *Gobi-Alt.* (Gurban-Bogdo and Gurban-Saikhan mountain ranges); *Alash. Gobi* (Alashan mountain range, cent. and west. parts, alpine meadows, March 5, 1873—Przew.; Yamata gorge, nor.-west. and nor.-east. slopes, upper belt, June 13, 1908—Czet.).

IB. Kashgar: *West.* (nor. slope of Kunlun, Ak-Këz pass, 110–120 km south of Kargalyk on Tibet highway, south. rocky slope of trough of crossing, under juniper grove, June 5, 1959—Yun. and I.-F. Yuan')

IC. Qaidam: *hilly* (Sarlyk-Ula, nor. slope, Elisten-Kuku-Bulak area, 3300 m, near rocks, May 11, 1895—Rob.).

IIA. Junggar: *Tarb.* (Saur mountain range, south. slope, Karagaita valley, right bank creek valley of Bain-Tsagan, subalp. meadow, June 23, 1957—Yun. et al.); *Jung. Alatau* (?), *Tien Shan.*

IIIA. Qinghai: *Nanshan* (South Tetung mountain range, forest belt, in valley, May 12; same locality, in precipices, occasional, May 24—1873, Przew.; nor. slope of mountain range in alpine belt, 3050–3650 m, in precipices, July 31, 1880—Przew.; Tashita river valley, upper boundary of spruce forest, 2900 m, May 28, 1886—Pot.).

IIIB. Tibet: *Chang Tang* (nor. slope of Kerii mountain range, Kurab river gorge, 3950 m, under rocks in moist places, July 6, 1885—Przew.; in Kazvan pass [Katszyvan] along Sinkiang-Tibet highway, 3450 m, No. 539, June 5, 1959—Lee and Chu (A.R. Lee (1959)), *Weitzan* (left bank of Dichu [Yangtze] in cent. portion, on rocks, 3950 m, June 23, 1884—Przew.; Dzhagyn-Gol, 3950–4250 m, moist sandy banks, July 3; Donra area on Khichu river, 3950 m, in rocks and among scrub, July 16—1900, Lad.); *South.* ("Height above source of Tsangpo, 5015 m, July 13, 1907—S. Hedin"—Ostenf. and Pauls. l.c.).

IIIC. Pamir ("Mustag-Ata, rochers sud du Roung-Koul, 3960 m, July 13, 1906, Lacoste"—Danguy, l.c.; under Ucha pass, along clayey-rocky descent, June 17; Ulug-Tuz gorge in Charlysh river basin, on descent, June 26—1909, Divn.; Kashkasu river source, at 4200–5500 m alt, July 5; upper Kailyk river, at 4500–5000 m alt., alpine tundra, July 14—1942, Serp.).

General distribution: Jung.-Tarb., Nor. and Cent. Tien Shan, East. Pam.; Arct., Europe, Balk.-Asia Minor, Caucasus (nor.), Mid. Asia (Pam.-Alay, West. Tien Shan), West. and East. Sib., Far East, Nor. Mong. (Fore Hubs., Hent., Hang.), China (Dunbei, Nor.), Himalayas, Japan, Nor. Amer.

2. **L. tibetica** Baker ex Oliver in Hook. Ic. pl. 23 (1892) tab. 2216; Franch. in J. Bot. (Paris) 12 (1898) 193 cum var. *purpurascen* Franch. and var. *lutescens* Franch.; Forbes and Hemsley, Index Fl. Sin. 3 (1903) 140; Rehd. in J. Arn. Arb. 14 (1933) 6 (var. *purpurascens*); Walker in Contribs. U.S. Nat. Herb. 28 (1941) 603 (var. *purpurascens*); Ching in Bull. Fan Memor. Inst. Biol. (Bot.) 10 (1941) 261. —Ic.: Hook. Ic. pl. tab. 2216.

Described from South-west China (Sichuan). Type in London (K). Plate V, fig. 5.

Exposed meadow slopes and wet rocks in alpine belt.

IA. Mongolia: *Alash. Gobi* (Alashan: "Ha La Hu Kou, No. 54, Ho Lan Shan, No. 1148", 1923, R.C. Ching—Walker, l.c., Ching, l.c.).

IIIA. Qinghai: *Nanshan* ("Ta Pan Shan, No. 675", 1923, R.C. Ching—Walker, l.c.); *Amdo* ("Alpine region between Radja and Jupar ranges, June 1926, J. Bock"—Rehd. l.c.).
General distribution: China (South-west).

13. Asparagus L.
Sp. pl. 1 (1753) 313; Gen. pl. (1754) 147.

1. Stem simple, not branched, rarely with 1–2 short branches in lower part, erect. Cladodes linear, up to 50 mm long, rather flat, glabrous, falcate upward. Rhizome long, funiform ....
.... 6. **A. przewalskyi** Ivanova.

+ Stem highly branched. Rhizome thick, shortened, densely covered above with remnants of dead stems and below with thick funiform roots .... 2.

2. Stem climbing or creeping, flexuous, long .... 3.

+ Stem erect, straight or geniculate. .... 4.

3. Cladodes in filiform dense clusters of 20–50 preserved until end of vegetative phase throughout stem length at base of first order branches. Cladodes slender, glabrous like stem and branches. Pedicels 10–20 mm long, jointed just below flower ....
.... 5. **A. neglectus** Kar. et Kir.

+ Cladodes absent at base of first order branches; if present, caducous. Cladodes linear or cylindrical-linear, rigid and prickly, 5–10 (20) mm long, in clusters of 4–10. Stem, branches and cladodes usually chondroid-tuberculate to some degree, rarely glabrous; cladodes often glabrous, but branches tuberculate-striate. Pedicels 5–10 (15) mm long, with joint in upper half ....
.... 7. **A. trichophyllus** Bge.

4. Cladodes slender, about 0.5 mm thick, 15–50 mm long, poorly divaricate. Stem usually single, straight, 25–75 cm tall. Pedicels longer than perianth of staminate flowers. Plant bright green .....
.... 4. **A. dahuricus** Fisch.

+ Cladodes more or less thick, rigid, 1–1.5 mm thick, highly divaricate, like branches, diverging nearly at right angle or even deflexed. Stems geniculately curved. Flowers sessile or pedicellate, pedicels not longer than staminate flowers .... 5.

5. Stem more often single, 30–60 mm tall. Cladodes 20–40 mm long, usually deflexed, glabrous like stem and branches. Middle scale-like cauline leaves with 2 mm long spur. Berries bright red up to tip, paired on main stem and branches. Plant apple-green ....
.... 1. **A. angulofractus** Iljin.

+ Stems usually 2–5, 10–30 cm tall. Cladodes 5–15 mm long, diverging at right angle or nearly so. Mature berries black, paired on main stem. Plants greyish-green .... 6.
6. Stem yellowish below. Branches and cladodes chondroid—finely tuberculate. Middle cauline leaves without spur ....
.... 2. **A. breslerianus** Schult.
+ Stem with white, flaking periderm in lower half. Branches and cladodes more or less glabrous. Middle cauline leaves with spur about 1 mm long. .... 4. **A. gobicus** Ivanova.

1. **A. angulofractus** Iljin in Fl. SSSR, 4 (1935) 746, 432; Fl. Kazakhst. 2 (1958) 222. —*A. soongoricus* Iljin in Fl. SSSR, 4 (1935) 747, 432; Fl. Kazakhst. 2 (1958) 223. —*A. ledebourii* auct. non Miscz.: Kryl. Fl. Zap. Sib. 3 (1929) 650; Pazij in Opred. rast. Sr. Azii [Key to Plants of Mid. Asia], 2 (1971) 117. —*A. inderiensis* auct. non Blum: Kryl. Fl. Zap. Sib. 3 (1929) 650. —*A. maritimus* auct. non Pall.: Ostenf. and Pauls. in Hedin, South. Tibet, 6, 3 (1922) 90? —*A. verticillatus* auct. non L.: Rgl. in Acta Horti Petrop. 6, 2 (1880) 535, quoad pl. e vall. Ili prope Kuldscha. —Ic.: Fl. SSSR, 4, Plate XXV, figs. 1 and 5; Fl. Kazakhst. 2, Plate XX, figs. 2 and 4.

Described from East. Kazakhstan (Ili river valley). Type in Leningrad.

Pebble and rubble deserts, saline sand, exposures of Tertiary (Gobi) rocks, debris cones and arid pebble beds.

IA. **Mongolia:** *Alash. Gobi* (10 km nor. of Bayantaokhai settlement [nor. of Inchuan' town], rocky-rubble plain, July 12, 1957—Petr.); *Khesi* (from Chzhae to Gaotai, No. 405, May 18; from Tszyutsyuan' to Yuimyn', No. 417, May 19—1957, Kuan).

IB. **Kashgar:** *Nor.* (Karateke, Taushkandar'ya about 1500 m from old cultivated fields, June 3, 1889—Rob.; between Aksu and Kuchei near Karayulgan settlement, Aug. 12, 1929—Pop.); *West.* (Yarkenddar'i basin, Chup river, Aug. 10, 1890—Grombch.; near Ulugchat town, dry red low hillocks, July 2; Sarykol'sk mountain range, Bostan-Terek area, July 12—1929, Pop.; 75 km nor.-west. of Kashgar on highway to Torugart, 2240 m, rubble valley trail, June 18; 83 km nor.-west of Kashgar along highway to Baikurt settlement, 2300 m, hilly steppised desert, along gully, June 19; 10–12 km nor. of Baikurt settlement along highway, desert belt, debris cone, June 20—1959, Yun. and I.-F. Yuan'; along Ulugchat-Baikurt road, 2200 m, No. 09705; same locality, 2300 m, along edge of irrigation ditch, No. 09708—June 19, 1959, Lee et al. (A.R. Lee (1959)); *South.* (Polur settlement, July 5, 1890—Grombch.; 25 km from Polur along road to Keriya, sympegma desert, May 10, 1959—Yun. and I.-F. Yuan'; Kun'-lun' foothills along Keriya-Polur road, 2000 m, No. 00059, May 10, 1959—Lee et al. (A.R. Lee (1959)).

IIA. **Junggar:** *Zaisan* (Ch. Irtysh river, left bank, Maikain area, hummocky sand June 7, 1914—Schischk.); *Dzhark.* (Bayandai near Kul'dzha, July 11; Suidun town, July 16—1877; Suidun town, May 8, 1878—A. Reg.).

General distribution: Fore Balkh.

Note. The name *A. ledebourii* Miscz. is inapplicable to this species since P. Misczenko described his species on the basis of K.A. Meyer's specimens from the vicinity of Baku (type in Leningrad) which clearly belong to *A. persicus* Baker. The reference of P. Misczenko to the rather imprecise flower *A. maritimus* drawn by Ledebour (Ledeb. Ic. pl. fl. ross. 4, tab. 393; hence the name *A. ledebourii* !) is erroneous since Ledebour described and depicted *A. maritimus* from Kazakhstan specimens (solonchaks along Irtysh near Arkaul mountains) and not from Transcaucasus as presumed by Misczenko.

2. **A. breslerianus** Schult. in Roem. et Schult. Syst. Veg. 7, 1 (1829) 323; Baker in J. Linn. Soc. (London) Bot. 14 (1875) 598; Miscz. in Monit. Jard. Bot. Tiflis, 12 (1916) 33; Iljin in Fl. SSSR 4 (1935) 431; Fl. Kazakhst. 2 (1958) 222; Pazij in Opred. rast. Sr. Azii [Key to Plants of Mid. Asia], 2 (1971) 114. —Ic.: Fl. SSSR, 4, Plate XXV, fig. 4; Fl. Kazakhst. 2, Plate XX, fig. 1.

Described from Fore Asia. Type in München (M) ?

Solonchaks, saline sand, solonetz clayey-rubble desert trails and foothill slopes.

IIA. **Junggar.** *Jung. Gobi* (nor.: Ulyungur lake near Salburta [cape], Aug. 16, 1876—Pot.); *Zaisan* (Ch. Irtysh, right bank below Burchum river estuary, Sary-Dzhasyk, Kiikpai well, solonchak, June 15; left bank of Ch. Irtysh opposite Cherektas mountain, tugai, June 11—1914, Schischk.).

General distribution: Aralo-Casp., Fore Balkh.; Fore Asia (Iran), Caucasus (east. and south. Transcaucasus), Mid. Asia (plains).

Note. The cited specimens are defective, without flowers or fruits and their identification thus rather tentative.

3. **A. dahuricus** Fisch. ex Link, Enum. pl. Horti Berol. Alt. 1 (1821) 340; Baker in J. Linn. Soc. (London) Bot. 14 (1875) 599; Franch. Pl. David. 1 (1884) 301; Forbes and Hemsley, Index Fl. Sin. 3 (1903) 101; Danguy in Bull. Mus. nat. hist. natur. 20 (1914) 138; Iljin in Fl. SSSR, 4 (1935) 433; Kitag. Lin. Fl. Mansh. (1939) 133; Grub. Konsp. fl. MNR (1955) 95. —*A. gibbus* Bge. Enum. pl. China bor. (1832) 65; Baker in J. Linn. Soc. (London) Bot. 14 (1875) 600; Forbes and Hemsley, l.c. 102; Kitag. Lin. Fl. Mansh. (1939) 133; Palibin in Tr. Troitskosavsko-Kyakht. otdelen. RGO, 7, 3 (1904) 42. —*A. tuberculatus* Bge. ex Iljin in Fl. URSS, 4 (1935) 747 and 433; Grub. Konsp. fl. MNR (1955) 96.

Described from East. Siberia (Dauria). Type in Leningrad.

Steppe and arid meadow, as well as rubble and rocky slopes, sandy steppes, fringes of pine groves, among steppe thickets.

IA. **Mongolia:** *Cen. Khalkha, East. Mong., Val. Lakes* (south. slope of Hangay), *East. Gobi* (nor.), *Ordos* (east.).

General distribution: East. Sib. (Daur.), Far East (Zee-Bur.), Nor. Mong. (Hent., Mong.-Daur.), China (Dunbei south., North).

Note. Separation of *A. tuberculatus*, like *A. gibbus* Bge., was irrational and M.M. Iljin (l.c.) described this species in vain. The extent of tuberculation of cladodes and branches varies greatly throughout the distribution range of the species from distinctly tuberculate all over to altogether glabrous, and is related neither to number of cladodes nor other characteristics. Glabrous as well as tuberculate specimens are found in a given population and even the same plant may exhibit some tuberculate branches and some totally glabrous ones.

98

4. **A. gobicus** Ivanova ex Grub. in Not. syst. (Leningrad) 17 (1955) 9; Grub. Konsp. fl. MNR (1955) 95.

Described from Mongolia (Valley of Lakes). Type in Leningrad. Plate V, fig. 2; map 4.

Desert, desert and sandy steppes on thin sand, sandy-pebble and sandy-rocky soils along intermontane valley floors and hill trails.

IA. **Mongolia:** *Mong. Alt.* (along slopes of mountains encircling gorge, leading from Sudzhi spring to valley running along foot of Burkhan-Ula mountain range [Beger-Nur valley], Aug. 21, 1894; mountains between Tsagan-Usu and Senkir river [Tsenkher], to south-east of Kobdo, July 17, 1898—Klem.; Khasagtu-Khairkhan mountain range, west. slopes of Undur-Khairkhan hills, on rock talus, Sept. 16; same site east. rocky slopes of Kukhengir mountains, Dundu-Tseren-Gol river, Sept. 18—1930, Pob.; Tonkhil somon, gorge, highly rocky slope, July 16, 1947—Yun.; east. extremity of Shoroitiin-Nuru near Ortsak-Ul along road to Delger somon from Bain-Tsagan somon, below rocks, Aug. 29; Khan-Taishiri mountain range, 5 km south-west of Dzun-Bulak along road, feather-grass-onion-snakeweed steppe on rather thin sand, Sept. 3—1948, Grub.; Tugurik-Gola gorge near estuary, left bank, pebble bedrock floor with boulders, June 27 1971—Grub. et al.); *Cen. Khalkha* (15 km nor. of Erdeni-Dalai somon on old Ulan Bator-Dalan Dzadagad road, feather grass-onion-snakeweed steppe, Aug. 13, 1948—Grub.); *Depr. Lakes* (Ubsi depression, 7 km from Khirgiz-Nur lake, on nor. bank of Baga-Nur, July 31, 1879—Pot.; between Kurginein-Khuduk and Dolon-Tur [Shargain-Gobi], sandy steppe, Aug. 21, 1896—Klem.; 40 km nor. of Kobdo along road to Tsagan-Nur, grass-pea shrub desert steppe on loam, Aug. 7, 1945—Yun.; south. Khara-Usu-Nur basin environs along road to Kobdo, feather-grass-biurgun desert, June 27, 1971—Grub. et al.; right bank of Khungui, Boro-Khara-Elesu sand along road to Urgomal somon, feather-grass-wormwood desert, Aug. 18, 1972—Grub. et al.); *Val. Lakes* (on right bank of Tuin-Gol, in arid creek valley on rock talus, Aug. 9; not far from Orok-Nur lake, pebble bed, Aug. 13, 1893; on descent to Baidarik river not far from road from Kobdo, Kuku-Khoto, semidesert, July 19, 1894—Klem.; mountains along Tuin-Gol river, dry rock bed, April 2, 1924—Pavl.; Ongin-Gol, sand-dunes, July 18, 1926—Lis.; plain south of Karaganat mountains and nor. of Argalinte mountains, Aug. 3, 1926—Glag.; Dzhinsetu somon, Barun-Khongor area, steppe on highly sandy soils, Aug. 17, 1949, Barun-Bayan-Ulan somon, plain on left bank of Tatsiin-Gol, feather-grass steppe, July 26, 1951—Kal.; Bilgekhin-Dzun-Kholyun area 20 km nor.-west of Tugrik somon centre, hummocky solonchak-like sand on lowland

floor, Aug. 23, 1952—Davazamd; Tatsiin-Gol and Tuin-Gol interfluvine region in Abzog-Uly area and Adagiin-Khara-Khuduk, Lygaryn-Ama, sandy trail of canyon floor, June 13, 1971—Grub. et al.); *Gobi-Alt.* (Bain-Ula mountain range, east. slope, lower belt, on sand, May 7; Dundu-Saikhan mountain range, south. valley, on sand, July 2, 1909—Czet.; Artsa-Bogdo range, shaly placer, July end, 1922—Pisarev; Legin-Gol valley near Leg somon, Aug. 19—Sept. 1; Nemegetu-Nuru, south. trail, in sandy gorge, Sept. 9—1927, M. Simukova; Bain-Tsagan mountains, creek valley, on dry bed, Aug. 9; Tszolin mountains, in creek valley, on rubble soil, Sept. 8—1931, Ik.-Gal.; Bain-Leg somon, Khatyn-Suudul-Ula, Sept. 7, 1970—Emel'yanov); *East. Gobi* (Tsagan-Tugurik—Getszygen-Gashun, June 4, 1831—I. Kuznetsov; on way to Urgu from Alashan, second half of July to first half of Aug. 1873—Przew.; along road to Urgu from Alashan from Tsat-Kholun well to Kharmyktai well, July 2, 1909—Czet.; Shabarakh usu, dry hills at 3700 ft., 1925—Chaney; Baga-Ude, rocky slopes of Khara-Ula mountains, Aug. 11 and 14; Baga-Ude, near Urguni-Ulan well, narrow sandy strip, Aug. 15—1926, Lis.; Ulan-Khairkhan upland west of Ude, Aug. 17, 1928

—L. Shastin; Khan-Bogdo somon, Khoir-Ul'dzeitu area near Khutag-Ula, desert steppe, Aug. 25; Delger-Hangay somon, Khoir-Ul'dzeitu and Sharangad areas, desert steppe, Sept. 8–15—1930, E. Kuznetsov; Kalgan road between Sain-Usu and Udzyr-Ulan-Dzhul'chi wells, sandy-pebble semidesert, Aug. 18; same site, Udzyr-Ulan-Dzhul'chi well, on bank of dry sandy bed of Motne-Gol, Aug. 19; same locality, Udzyr-Ulan-Dzhul'chi well, along slopes of rather low ridges, Aug. 19; same locality, near Baga-Ude well, along slopes and in Bain-Gote mountain valleys, Aug. 21; same locality, road between Dolo-Chulut basin and Ude well, semidesert, Aug. 24; 17 km from Dzamyin-Ude custom house, Khukh-Tologoi well, rubble semidesert, Aug. 27; 25 km nor.-east of Argaleul hills, in clayey basin, Sept. 1—1931, Pob.; 16 km nor.-east of Sain-Shandy along road to Baishintu, chee grass thickets on gully floor, June 16, 1940—V. Shubin; 40 km south-east of Dalan-Dzhirgalan somon, Delgeriin-Deris area, feather-grass-onion steppe Aug. 24; 10 km nor.-east of Khara-Airik somon along road to Choiren, lowland solonchak steppised desert, Aug. 26; Khara-Airik somon, 20–23 km south-east of Buterin-Obo on ulan-Bator—Sain-Shanda road, feather-grass-wormwood desert steppe, Aug. 27—1940, Yun.; 20 km nor. of Mandal-Obo somon, on slope of mud cone, May 8, 1941—Yun.; Delger-Khid somon, Bosatyn-Tala area, pea shrub-feather-grass steppe, June 2, 1941—I. Tsatsenkin; Undur-Shili somon, 10 km south-east of Notyin-Ulan-Nur lake, feather-grass-tansy desert steppe, June 2; Undur-Shili somon, slope of Kholbo-Obo knoll to Undegin-Toirim lowland, feather-grass desert steppe along trail, June 2; 40 km east of Khara-Airik somon, pea shrub-feather-grass steppe, June 4; 25 km west of Ikhe-Dzhirgalan somon on road to Bayan-Dzhirgalan, smoothened hummocky area, June 31; 32 km nor.-east of Dzamyn-Ude along road to Baishintu, feather-grass desert steppe on ridge trail, June 12; 38 km nor.-west of Dzamyn-Ude, pea shrub-feather-grass desert steppe, June 12; Erdeni somon, Borokha-Tala area, tansy-feather grass desert steppe, June 16—1941, Yun.; Bain-Dzak area, precipices of thin sand Gobi beds, Sept. 5, 1950—E. Lavrenko and Yun.; 70 km from Baishintu along road to Sain-Shandu, Aug. 26; 125 km from Sain-Shandy along road to Ulan-Bator, Sept. 5—1950, A.T. Ivanov; Bailinmyao, desert steppe, 1959—Ivan.); *Alash. Gobi* (Dyn'yuan'in oasis on clayey soil, June 3: Tengeri sand, Dolone-Gol area, sandy-clayey soil, Aug. 13 1908—Czet.; Noyan somon, 20 km south of Oldakhu-Khita, desert steppe, June 7, 1949—Yun.; 15 km west of Bayan-Khoto, partially overgrown sand, Aug. 5, 1957; 25 km south of Bayan-Khoto, west. submontane pebble plain of Alashan mountain range, June 10, 1958—Petr.; Bordzon-Gobi in Ulan-Tsab precipice area, along flank of gorge, July 29, 1970—Grub. et al.); *Ordos* (15 km south of San'shingun town, right bank of Huang He, high sandy-pebble terrace, June 7; 40 km south of Dynkou town, valley in Alabusushan mountains, pebble-sandy slopes, June 8—1958, Petr.); *Khesi* (Loukhushan' mountain range, Aug. 17, 1908—Czet.; 67 km nor. of Lan'chzhou town, 1930 m, semidesert, June 21; Beitashan mountain nor. of Lan'chzhou town, dry slopes of loessial mounds, June 24—1957, Petr.; Gaolan', Kheishan pits, low-hill precipices, June 22; 15 km nor. of Yunchan town, rocky slopes of Beidashan mountains, June 28—1958, Petr.).

IIA. Junggar: *Jung. Gobi* (east.: 12 km south-west of Altay-Gasani-Uly along road to Bodonchin-Baishing, desert steppe on rather thin sand, Sept. 10, 1948—Grub.).

General distribution: endemic.

Note. The species is very closely related to Kazakhstan-Fore Asian *A. breslerianus* Schult. and has similar ecology.

5. **A. neglectus** Kar. et Kir. in Bull. Soc. natur. Moscou, 14 (1841) 48; Iljin in Fl. SSSR, 4 (1935) 430; Fl. Kirgiz. 3 (1951) 118; Fl. Kazakhst. 2 (1958) 222; Fl. Tadzh. 2 (1963) 281; Pazij in Opred. rast. Sr. Azii [Key to Plants of

Mid. Asia], 2 (1971) 115.—*A. flexuosus* Kryl. in Animadv. Syst. Herb. Univ. Tomsk, 9 (1928) 1; Kryl. Fl. Zap. Sib. 3 (1929) 647, p.p. —Ic.: Fl. SSSR, 4, Plate 25, fig. 11.

Described from South. Altay. Type in Leningrad.

Among shrubs on river-banks, shady ravines and creek valleys, rock talus, rock shade and thickets on slopes.

IIA. Junggar: *Cis-Alt.* (Shara-Sume settlement [Altay settlement], 1100 m, No. 2411, Aug. 26, 1956—Ching; along Shara-Sume—Karamai road, in ravine, No. 10792, July 30, 1959— Lee et al. (A.R. Lee (1959)); *Jung. Alatau* (10 km nor. of Toli town, on slope, No. 4970, Aug. 4, 1957—Kuan); *Tien Shan* (Urtas-Aksu, June 17, 1878—Fet.; Khorgos river [May 16] 1878; Tsagan-Tyunge, 1500–1800 m, June 8, 1879—A. Reg.; nor. slope, Usu district, San'tszyaochzhuan' village, in wet area, No. 1030, June 25, 1957—Kuan; Ketmen' mountain range, 3–4 km beyond Sarbushin settlement along road to Kzyl-Kure from Kul'dzha, steppe belt, on steep talus slope, Aug. 23, 1957—Yun. et al.); *Jung. Gobi* (nor.: [crossing] Dyurbel'dzhin [along Ch. Irtysh river], meadow, Aug. 24; Ch. Irtysh river, Aug. 26—1876, Pot.; south: bank of Manas river, desert, No. 789, June 11, 1957—Kuan); *Zaisan* (Lasta river, Aug. 7; along road to Tumanda river from Lasta river, Aug. 7—1876, Pot.).

General distribution: Aralo-Casp. (east.: Ulutau), Fore Balkh., Jung.-Tarb., Nor. Tien Shan; Mid. Asia (West. Pam.-Alay), West. Sib. (Altay south.).

6. **A. przewalskyi** Ivanova sp. nova hoc loco (sect. *Asparagus*).

Planta dioica, humilis, 10–30 (raro ad 50) cm alta, rhizomate repente funiculiformi. Caulis erectus, laevis, simplex rarius in parte ima ramis brevibus evolutis, foliis squamiformibus ecalcaratis, cladodiis subaequalibus linearibus complanatis acutis plerumque 5–8 in fasciculo 1.5–3 (5) cm longis arcuatis. Flores in caule parte ima dispositi bini, pedicellis nutantibus perigonio aequilongis vel brevioribus sub flore articulatis, perigonio brunneo campanulato-infundibuliformi 7–8 mm longo. Baccae rubrae, in sicco ca. 7 mm diametro.

Typus: Tsinghai, systema fluvii Hoangho superiores, latter half of May, 1880, Przewalskyi. In Herb. Inst. Bot. Ac. Sci. URSS (Leningrad) conservatur.

Species propria caulo simplici vel parum ramoso, cladodiis complanatis nec non rhizomate longo funiculiformi ab aliis valde diversa. Plate V, fig. 1.

IIIA. Qinghai: *Nanshan* (North Tetung mountain range [near Chertynton temple], March 31, 1872; [South Tetung mountain range] near Cheibsen temple, in coniferous forest, July 28, 1880—Przew.); *Amdo* (basin of Huang He river upper course, latter half of May, 1880—Przew., typus!).

General distribution: China (North-West: Gansu, south of Lan'chzhou)

7. **A. trichophyllus** Bge. Enum. pl. China Bor. (1832) 65; Baker in J. Linn. Soc. (London) Bot. 14 (1875) 600; Rgl. in Acta Horti Petrop. 6, 2 (1880) 535, p.p.; Franch. Pl. David. 1 (1884) 301; Forbes and Hemsley, Index Fl. Sin. 3 (1903) 103; Kitag. Lin. Fl. Mansh. (1939) 134; Walker in Contribs. U.S. Nat.

Herb. 28 (1941) 603; Chen and Chou, Rast. pokrov r. Sulekhe [Vegetation Cover of Sulekhe River] (1957) 92. —*A. brachyphyllus* Turcz. in Bull. Soc. natur. Moscou, 13 (1840) 78; Baker in J. Linn. Soc. (London) Bot. 14 (1875) 602; Forbes and Hemsley, Index Fl. Sin. 3 (1903) 101; Iljin in Fl. SSSR, 4 (1935) 437; Kitag. Lin. Fl. Mansh. (1939) 133; Walker in Contribs. U.S. Nat. Herb. 28 (1941) 603; Fl. Kazakhst. 2 (1958) 223; Fl. Tadzh. 2 (1963) 285; Pazij in Opred. rast. Sr. Azii [Key to Plants of Mid. Asia], 2 (1971) 116. —*A. trichophyllus* γ. *trachyphyllus* Bong. et Mey. in Kunth, Enum. 5 (1850) 68. —*A. pallasii* Miscz. in Monit. Jard. Bot. Tiflis, 12 (1916) 27; Kryl. Fl. Zap. Sib. 3 (1929) 649. —*A. maritimus* Pall. Reise, 2 (1773) 329, non Mill.; Bge. Enum. pl. China bor. (1831) 65; Franch. Pl. David. 1 (1884) 301. —*A. tamariscinus* Ivanova ex Grub. in Not. Syst. (Leningrad) 17 (1955) 9; Grub. Konsp. fl. MNR (1955) 95. —Ic.: Fl. SSSR, 4, Plate 25, fig. 2.

Described from North China (Beijing vicinity). Type in Paris (P), isotype in Leningrad.

In Chee grass and reed groves, sasa solonchak, coastal solonchaks and saline sand among shrubs, tugais, gorge floors and clayey precipices of rivers and ravines, among rocks, banks of irrigation ditches, solonchak marshes in deserts and in desert and arid steppes.

IA. **Mongolia:** *Depr. Lakes* (Dzeren-Nor, on sandy soil overland with salt crust, Aug. 6; Tatkhen-Teli-Gol, along banks, Aug. 10; Khara-Usu lake, nor. of Dzun-Khairkhan, on light sand, Aug. 16; Khara-Usu lake, south. end, on light sand [hummocky] blowing on stems of chee grass, Aug. 17; Kundelen river [west of Ulangom], solonetz clayey soil, in chee grass scrubs, Sept. 20—1879, Pot.); East. Gobi (around Daban-Urto unit, on loam between rocks, June 6, 1831—I. Kuznetsov; Mongolia Chinensis, 1840—Kirillov; along mountains of south-east. Mongolia, 1871–1873—Przew.; Barun-Khurne-Khuduk south of Galba mountain, on sand with reeds and tamarisk, Sept. 25, 1940—Yun.); *Transalt. Gobi* (Tsagan-Derisu, in tamarisk shrubs on sandy soil, July 3, 1877—Pot.; Shara-Khulusnii-Bulak, in poplar forest, July 27, 1973—Isach. and Rachk.); *Alash. Gobi* (Alashan, first half of June, 1872—Przew.; Edzin-Gol valley [near upper Ontsin-Gol], Bukhan-Khub area, sandy soil, June 3, 1926—Glag.; 11 km west of Mukhur-Shanda spring along border road south of Tostu-Nuru, low reed thickets and hummocky chee grass thickets, Aug. 14, 1948—Grub.; 15 km south-east of Minchin town, around Gaotszyasavo village, sand-covered solonchak-like meadow, July 3, 1958—Petr.; sandy Tengeri desert around Ivanch [Chzhunvei], July 23, 1957—Petr.); *Ordos* (south of Shine-Sume monastery, sand-dunes, Sept. 10; south of Narin-Gol river, sand-dunes, Sept. 11—1884, Pot.); *Khesi* (Suchzhou town, Aug. 1889—Martin; 24 km south-east of An'si town, Tasykhe river valley, meadow solonchak, July 27, 1958—Petr.).

IB. **Kashgar:** *Nor., West., South., East.*

IC. **Qaidam:** *plain* (East. Qaidam, Aug. 1879—Przew.; nor. slope of Burkhan-Budda mountain range along Nomokhun-Gol river, 2800 m, on wet solonchak soil, Sept. 11, 1884—Przew.; upper gorge of Dzuliin-Gol, 3350 m, along bushes and bulrush, May 3; East. Qaidam [Khudo-Bure area], in bulrush bushes on solonchak, May 15; Baga-Qaidamin-Nor, about 3050 m, wet solonchak, June 12—1895, Rob.; nor.-east. Qaidam, Dalyn-Turgyn area, 2800 m, solonchak marsh, Aug. 5, 1901—Lad.).

IIA. Junggar: *Tien Shan, Jung. Gobi, Zaisan, Dzhark.*

IIIA. Qinghai: *Nanshan* (west. extremity of South Tetung mountain range, second half of April to first half of May, 1873—Przew.; Dangertin town vicinity, 2450 m, Aug. 12, 1901—Lad.); *Amdo* (Baga-Gorgi river, left tributary of Huang He, May 6; pass through Syan'sibei mountain range, May 28—1880, Przew.).

General distribution: Aralo-Casp., Fore Balkh.; Europe (south-east. European part of USSR), Mid. Asia (plains), West. Sib. (south), East. Sib. (Ang.-Sayan. south.).

Note. N.S. Turczaninov described *A. brachyphyllus* Turcz. 10 years after *A. trichophyllus* Bge. from P.E. Kirillov's collection, also from Beijing vicinity and like *A. pallasii* Miscz., it should be treated as a pure synonym of *A. trichophyllus.* Tuberculation of cladodes and branches, as in the case of *A. dahuricus* Fisch.=*A. tuberculatus* Bge. ex Iljin, varies in a very wide range, from totally glabrous to distinctly and uniformly finely tuberculate, not only throughout the distribution range of the species but also within the same population and even in one and the same specimen as can be seen in Bunge's isotypes among which finely tuberculate branches occur alongside glabrous ones in most cases. There is no justification whatsoever to separate these two closely related forms as distinct species. The number of cladodes in the cluster, their length, thickness and "divarication" greatly depend on direct ecological factors in which the specimen has grown and the weather conditions in that year. Specimens growing in shade are usually more slender and soft, with numerous filiform cladodes (authentic A. Bunge's specimens fall in this category) while specimens from open sunny sites are rougher with short and rigid tuberculate cladodes (authentic specimens of N.S. Turczaninov). Plants from solonchaks and solonchak-like swamps are very thick, with thick divaricate cladodes. Plants with dimorphic branches are found on sand in different parts of the distribution range (for example, in Ordos and Kashgar). Some branches (more often lower ones) among these specimens have long and glabrous cladodes while others have short and tuberculate branches or clusters of slender glabrous cladodes occupying the upper portion of the branches and short and tuberculate ones the basal part. We should also accept the view of M.M. Iljin (Flora SSSR, 4: 437) that it is very difficult, if not impossible, to distinguish *A. trichophyllus* Bge. from *A. persicus* Baker.

14. Smilacina Desf.
in Ann. Mus. hist. natur. (Paris) 9 (1807) 52.

1. **S. trifolia** (L.) Desf. l.c.; Danguy in Bull. Mus. nat. hist. natur. 20 (1914) 138; Kuzeneva in Fl. SSSR, 4 (1935) 452; Kitag. Lin. Fl. Mansh. (1939) 142.—*Convallaria trifolia* L. Sp. pl. (1753) 316. —**Ic.:** Fl. SSSR, 4, Plate XXVII, fig. 2.

Described from Siberia. Type in London (Linn.).

Humid marshy meadows, humid willow and poplar inundate groves.

IA. Mongolia: *East. Mong.* ("Vallee du Keroulen, 1896, Chaff."—Danguy, l.c.).

General distribution: West. Sib. (south-east.), East. Sib., Far East, China (Dunbei), Korean peninsula, Nor. Amer.

15. **Majanthemum** Wigg.

Prim. Fl. Holsat. (1780) 14, nom. cons.

1. **M. bifolium** (L.) F.W. Schmidt, Fl. Boëm. inch. 4 (1794) 55; Danguy in Bull. Mus. nat. hist. natur. 20 (1914) 138; Kryl. Fl. Zap. Sib. 3 (1929) 651; B. Fedtsch. in Fl. SSSR, 4 (1935) 435; Kitag. Lin. Fl. Mansh. (1939) 140; Walker in Contribs. U.S. Nat. Herb. 28 (1941) 603; Grub. Konsp. fl. MNR (1951) 96; Fl. Kazakhst. 2 (1958) 227. —*M. convallaria* Wigg. Prim. Fl. Holsat. (1780) 15; Forbes and Hemsley, Index Fl. Sin. 3 (1903) 112. —*Convallaria bifolia* L. Sp. pl. (1753) 316. —*Smilacina bifolia* Schult. f. Syst. veg. 1 (1829) 307; Franch. Pl. David. 1 (1884) 303. —**Ic.**: Reichb. Ic. Fl. Germ. 10, tab. 336; Fl. SSSR, 4, Plate XXVII, fig. 4.

Described from Nor. Europe. Type in London (Linn.).

Coniferous and mixed forests, birch groves and forests, thickets.

IA. Mongolia: *East. Mong.* (Muni-Ula, nor. slope, in scrubs on mountains slopes, humus, June 25, 1871—Przew.; "Vallee du Keroulen, 1896, Chaff."—Danguy, l.c.); *Alash. Gobi* (Alashan mountain range near Baisa monastery, 2500 m, spruce-pine forest, July 6, 1957—Petr.).

IIIA. Qinghai: *Nanshan* (South Tetung mountain range, 2300 m, on humus soil, July 26, 1880—Przew.; "Tu Er Ping, in woods, June 1923, R.C. Ching"—Walker, l.c.).

General distribution: Europe, Mediterr., Balk., West. and East. Sib., Far East, Nor. Mong., China (Dunbei, North, North-West), Korean peninsula, Japan.

16. **Streptopus** Michx.

Fl. bor. amer. 1 (1803) 200.

1. **S. amplexifolius** (L.) DC. in Lam. and DC. Fl. France, 3·(1805) 174; Maxim. in Bull. Ac. Sci. St.-Pétersb. 29 (1883) 212; Forbes and Hemsley, Index Fl. Sin. 3 (1903) 110; B. Fedtsch. in Fl. SSSR, 4 (1935) 455; Walker in Contribs. U.S. Nat. Herb. 28 (1941) 603. —*Uvularia amplexifolia* L. Sp. pl. (1753) 304. —**Ic.**: Michx. l.c. tab. 18; Reichb. Ic. Fl. Germ. 10, tab. 421, fig. 959; Fl. SSSR, 4, Plate XXVII, fig. 5.

Described from Europe. Type in London (Linn.).

In humid mossy forests and scrubs.

IIIA. Qinghai: *Nanshan* (South Tetung mountain range, lower mountain belt, mossy deciduous forest, in clusters, common, Aug. 11, 1872; same locality, forest belt, under rocks in humus soil, 2600 m, Aug. 5, 1880—Przew.; "Tai Wang Kou, at base of rocky cliff in a forest, June 1923, R.C. Ching"—Walker, l.c.).

General distribution: Cent. Europe, Mediterr., East. Sib. (Leno-Kol.), Far East, China (Dunbei, North, North-west, South-west), Japan.

Note. Nanshan specimens cited by us are small—not more than 25 cm tall—with unbranched stem; moreover, some have slender long offshoots emerging from rhizome. These characteristics are not exclusive to Nanshan plants but also reported from different parts of the distribution range of the species. In all other characteristics, Nanshan plants are indistinguishable from typical ones, including the characteristic structure and arrangement of pedicels, and hence we fully endorse the view expressed by K.I. Maximowicz (l.c.) that there is no rationale for taxonomic differentiation of these plants.

17. **Polygonatum** Adans.
Fam. pl. 2 (1763) 53.

1. Leaves alternate .... 2.
+ Leaves in whorls and partly alternate .... 3.
2. Stem angled. Peduncles 1–1.5 cm long, 1–2-flowered; perianth tubular, without constriction; filaments of stamens glabrous ......
.... 4. **P. odoratum** (Mill.) Druce.
+ Stem cylindrical. Peduncles 3–5 cm long, 3–5-flowered; perianth campanulate-tubular, narrowed above ovary, but enlarged on top; filaments of stamens pilose .... 3. **P. macropodum** Turcz.
3. Flowers pink. Leaves not curled at tip, acute .... 4.
+ Flowers white or greenish-brown. Leaves curled at tip .... 5.
4. Flowers in pairs on bifurcate peduncles. Leaves 7–15 cm long and 1–2 cm broad; lower ones in whorls of 4 ....
.... 6. **P. roseum** (Ledeb.) Kunth.
+ Flowers solitary. Leaves 5–7 cm long and 5–12 mm broad, lower ones in whorls of 3 .... 2. **P. kansuense** Maxim.
5. Leaves glabrous beneath. Flowers white ....
.... 7. **P. sibiricum** Delaroche.
+ Leaves puberulent beneath along veins .... 6.
6. Flowers greenish-brown, on long bifurcate peduncles, 2–4 each in axil. Stem up to 1.5 m tall. Leaves lanceolate, up to 2 cm broad, in whorls of 3–5–8 .... 1. **P. fuscum** Hua.
+ Flowers pinkish or white, on short bifurcate or simple peduncles, a few together in axil. Stem 20–50 cm tall. Leaves linear, 3–10 mm broad, in whorls of 3–6 .... 5. **P. prattii** Baker.

1. **P. fuscum** Hua in J. Bot. (Paris) 6 (1892) 444; Forbes and Hemsley, Index Fl. Sin. 3 (1903) 105. —*P. bulbosum* Lévl. in Feddes repert. 11 (1912) 302; Rehder in J. Arn. Arb. 14 (1933) 6. —? *P. sibiricum* auct. non Delaroche-Hao in Bot. Jahrb. 68 (1938) 588.

Described from China (Yunnan). Type in Paris (P).

Coniferous and deciduous forests.

IIIA. **Qinghai:** *Nanshan* (South Tetung mountain range, nor. slope, deciduous forest, Aug. 10, 1872—Przew.; Chorten-Ton temple, 2100–2400 m, spruce forest, Sept. 8, 1901—Lad.; Sining-fu, Shangwu-chuang, 2900 m"—Hao, l.c.); *Amdo* ("Radja and Yellow River gorges, leg. Rock"—Rehder, l.c.).

IIIB. **Tibet:** *Weitzan* (Mekong basin, along Dzechu river near Derchzhiling temple [96°40′ E. long. and 32°20′ N. lat.], 3500 m, spruce forest, Sept. 6, 1900—Lad.).

General distribution: China (North-west: Gansu; South-west).

2. **P. kansuense** Maxim. ex Batal. in Acta Horti Petrop. 11, 2 (1891) 493; Hua in J. Bot. (Paris) 6 (1892) 426; Forbes and Hemsley, Index Fl. Sin. 3 (1903) 105.

Described from China (Nanshan). Type in Leningrad.

Coastal and forest meadows.

IIIA. **Qinghai:** *Nanshan* (South Tetung mountain range, middle belt, grassy mountain slope, in clusters, common, May 13 [25], 1873, Przew.—typus!; Xining mountains, Myndan'sha river, June 1, 1890—Gr.-Grzh.).

General distribution: China (North-west: south-east. Gansu).

3. **P. macropodum** Turcz. in Bull. Soc. natur. Moscou, 5 (1832) 205; Maxim. in Bull. Ac. Sci. St.-Pétersb. 29 (1883) 208; Franch. Pl. David. 1 (1884) 302; Hua in J. Bot. (Paris) 6 (1892) 420; Forbes and Hemsley, Index Fl. Sin. 3 (1903) 106; Kitag. Lin. Fl. Mansh. (1939) 141. —*P. umbellatum* Baker in J. Linn. Soc. (London) Bot. 14 (1875) 553. —*P. multiflorum* auct. non All.: Ching in Bull. Fan. Memor. Inst. Biol. (Bot.) 10 (1941) 261; Walker in Contribs. U.S. Nat. Herb. 28 (1941) 603.

Described from Nor. China (Kalgan vicinity). Type in Leningrad.

Forests, among scrubs, in rock shadows.

IA. **Mongolia:** *Alash. Gobi* (Alashan mountain range: "Peissu kou, under bushes in a forest, common, May, 1923, R.C. Ching"—Walker, l.c.).

General distribution: China (Dunbei, North).

4. **P. odoratum** (Mill.) Druce in Ann. Scott. Nat. Hist. (1906) 226. —*P. officinale* All. Fl. Pedemont. 1 (1785) 131; Franch. Pl. David. 1 (1884) 302; Hua in J. Bot. (Paris) 6 (1892) 395; Forbes and Hemsley, Index Fl. Sin. 3 (1903) 107; p.p.; Danguy in Bull. Mus. nat. hist. natur. 20 (1914) 138; Kryl. Fl. Zap. Sib. 3 (1929) 653; Knorr. in Fl. SSSR, 4 (1935) 463; Grub. Konsp. fl. MNR (1955) 96; Fl. Kazakhst. 2 (1958) 229. —*Convallaria polygonatum* L. Sp. pl. (1753) 315. —*C. odorata* Mill. Gard. Dict. ed. 8 (1768) No. 4. —**Ic.:** Fl. SSSR, 4, Plate XXVIII, fig. 1.

Described from Europe. Type in London (BM).

Coniferous and deciduous forests, scrubs, shady crevices and cracks in granite rocks and outliers.

IA. **Mongolia:** *Cen. Khalkha* (Bichikte-Dulan-Khada massif, granites, Aug. 1, 1925—Gus.; Ikhe-Tukhum-Nur vicinity, Dulga mountain, June 1926—Zam.; Sorgol-Khairkhan-Ula 180 km south-east of Ulan-Bator along old road to Dalan-Dzadagad, along crevices between granite slabs, July 15, 1943—Yun.); *East. Mong.* ("Environs de Kailar, collines sablonneuses, 700 m alt., June 21, 1896, Chaff."—Danguy, l.c.; Khailar town, July 6, 1901—Lipsky; Khailar town, Sishan' mounds, lowland among sand-dunes, in forest, 600 m, Nos. 530 and 537, June 7, 1951—B. Skvortzov, T.-t. Li et al.; Khailar town, meadow steppe, 1959—Ivan.; Manchuria station vicinity, 1915, Nechaeva; Del'-Obo mountain on right bank of Kerulen river 6 km south of Khulun-Buir somon, scrubs on nor. slope, July 15, 1971—Dashnyam et al.; Suma-Khada mountains, June 4, 1871—Przew.); *Alash. Gobi* (Alashan mountain range, west. slope in midportion, forest, common, July 14, 1873—Przew.; same locality, Tszosto gorge, east. slope, lower belt, in scrubs, May 15; same locality, Dartymto gorge, in gully among scrubs, May 27—1908, Czet.).

IIIA. Qinghai: *Nanshan* (South Tetung mountain range, lower belt, common, May 28, 1873—Przew.).

General distribution: Europe, West. Mediterr., West. Sib., East. Sib. (south.), Nor. Mong., China (Dunbei, North).

5. **P. prattii** Baker in Hook. Ic. Pl. 23 (1892) pl. 2217; Hua in J. Bot. (Paris) 6 (1892) 420; Forbes and Hemsley, Index Fl. Sin. 3 (1903) 108. —*P. erythrocarpum* Hua, l.c. 424; Forbes and Hemsley, Index Fl. Sin. 3 (1903) 105. —*P. soulie* Hua in J. Bot. (Paris) 6 (1892) 427; Forbes and Hemsley, l.c. 109. —*P. sibiricum* auct. non Delaroche: ?Walker in Contribs. U.S. Nat. Herb. 28 (1941) 603. —? *P.* sp. Rehder in J. Arn. Arb. 14 (1933) 6.

Described from China (Sichuan). Type in London (K).

Forests, forest meadows, among scrubs, coastal pebble beds, clayey coastal cliffs.

IC. Qaidam: *mountains* (Dulan-Khit mountain, 3050 m, pebble bed, Aug. 10, 1901—Lad.).

IIIA. Qinghai: *Nanshan* ("Upper Shui Mo Kou, near Lien Cheng, in woods, rare; height 60 cm; flowers greenish-white, No. 395, July 1923, R.C. Ching"—Walker, l.c.); *Amdo* (along Mudzhik river south of Guidui town, April 16; along Churmyn river, on pebbles or clayey precipices in gorges, common, April 11; along Baga-Gorgi river, in forest, 2700 m, May 23—1880, Przew.; along Lanchzha-Lunva river, in meadows, May 14, 1885—Pot.).

General distribution: China (North-west: south. Gansu; South-west: Sichuan).

6. **P. roseum** (Ledeb.) Kunth, Enum. pl. 5 (1850) 144; Kryl. Fl. Zap. Sib. 3 (1929) 656; Knorr. in Fl. SSSR, 4 (1935) 458; Fl. Kirgiz. 3 (1951) 119; Fl. Kazakhst. 2 (1958) 228; Fl. Tadzh. 2 (1963) 289; Pazij in Opred. rast. Sr. Azii [Key to Plants of Mid. Asia], 2 (1971) 118. —*Convallaria rosea* Ledeb. Ic. pl. fl. ross. 1 (1829) 3, tab. 1; ej. Fl. alt. 2 (1830) 41. —**Ic.:** Ledeb. Ic. pl. fl. ross. 1, tab. 1; Curtis's Bot. Mag. 34, tab. 5049.

Described from South. Altay (Kurchum river). Type in Leningrad.

Sparse spruce forests, coastal meadows and scrubs, forest meadows, shady meadow slopes and rock shadows.

IIA. Junggar: *Tien Shan* (Khanakhai mountains, 1500–2100 m, June 16, 1878—A. Reg.; Burkhantau, June 5; in Naryn-Kola valley [Tekes tributary], 1700 m, Aug. 8—1878, Fet.; along Borotala river and at Chinese picket, Aug. 12, 1878—Larionov; Iren-Khabirga mountain range, Taldy river midcourse, 2100 m, May 25; Tsagan-Tyunge, 1500–1800 m, June 8; Aryslyn river, 3050–3350 m, July 19; Aryslyn estuary on Kash river, 1700 m, July 22—1879, A. Reg.; Tekes river and mountains, between scrubs on river, 3350 m, June 30, 1893—Rob.; bei Manass, June 5–17, 1908—Merzb.; in mountains south of Nyutsyuantsza, No. 652, July 18; Kizyl-Zangi village south of Shichan, on shaded slope, No. 792, July 23; same locality, on nor. slope, No. 4818, July 23; 6 km south-west of Shichan, 1700 m, on slope, No. 2259, July 26; hill south of Sin'yuan' town, near reservoir, No. 745, Aug. 22; same locality, 1900 m, on slope, No. 1144, Aug. 22—1957, Kuan; upper Tekes 4–5 km from Aksu settlement, nor. slope of main mountain range, forest belt, grassy spruce forest, Aug. 24; same locality, near bridge on Aksu—Kalmak-Kure road, in floodplain among scrubs, Aug. 25—1957, Yun. et al.).

General distribution: Jung.-Tarb., Nor. Tien Shan; Mid. Asia (Pam.-Alay), West. Sib. (Altay).

7. **P. sibiricum** Delaroche in Redouté, Liliac. 6 (1811) tab. 315; Franch. Pl. David. 1 (1884) 302; Forbes and Hemsley, Index Fl. Sin. 3 (1903) 108; Danguy in Bull. Mus. nat. hist. natur. 20 (1914) 138; Knorr. in Fl. SSSR, 4 (1935) 459; Kitag. Lin. Fl. Mansh. (1939) 141; Grub. Konsp. fl. MNR (1955) 96. —*P. fuscum* auct. non Hua: ?Walker in Contribs. U.S. Nat. Herb. 28 (1941) 603. —Ic.: Redouté, l.c.; Fl. SSSR, 4, Plate XXVIII, fig. 5.

Described from East. Siberia. Type in Geneva (G)?

Coniferous and deciduous forests, steppe scrubs, shady crevices and cracks in rocks and outliers, on stable sand.

IA. **Mongolia:** *Cen. Khalkha* (mid. Kerulen, slopes of Bain-Khan granite mountains; same locality, near Erdeni mountain—1899, Pal.; Tszakha mountain in Suchzh river valley, meadow slope, July 10, 1924—Pavl.; Choiren-Ula, feather-grass-wormwood steppe on gravelly trails of knolls, May 28; same locality, along crevices in granite outliers, May 28—1941, Yun.); *East. Mong.* (prope Dolon-Nor, 1870—Lom.; "Environs de Kailar, alt. 760 m, sables, June 24, 1896, Chaff."—Danguy, l.c.; Khailar town vicinity, in scrubs on hummocky sand, 600 m, No. 718, June 14, 1951—S.H. Li et al.; Manchuria station, Beishan' mountain, hummocky valley, No. 832, June 24, 1951—S.H. Li et al.; Shilin-Khoto town, true steppe, 1959—Ivan.; Dariganga, Daun-Nart upland, south-west. slope, basalt outcrops among scrubs, Aug. 10, 1970—Grub. et al.); *Alash. Gobi* (Alashan mountain range; west. slope midportion, forest, common, July 15, 1873—Przew.; Tszosto gorge, east. slope, midbelt, in thickets, May 9; same locality, along valleys in scrubs, May 15—1908, Czet.; "Ho Lan Shan, in woods, No. 1135; stem slender, up to 2 m tall, flowers white, May 1923, R.C. Ching"—Walker, l.c.); *Ordos* (sand-dunes nor. of Naryn-Gol river, in dense scrubs often in clusters, Sept. 10, 1884—Pot.; 30 km south of Dalat town, Huang Ho valley, Chzhandanchzhao village environs, meadow, Aug. 10, 1957—Petr.).

General distribution: East. Sib. (Daur.), Nor. Mong. (Hent., Hang., Mong.-Daur.), China (Dunbei, North).

19. Convallaria L.
Sp. pl. (1753) 314.

1. **C. keiskei** Miq. in Ann. Mus. Lugd.-Batav. 3 (1867) 148; Kitag. Lin. Fl. Mansh. (1939) 134; Grub. in Bot. zh. 56, 11 (1971) 1642. —*C. majalis* auct. non L.: Franch. Pl. David. 1 (1884) 303; Forbes and Hemsley, Index Fl. Sin. 3 (1903) 112; Knorr. in Fl. SSSR, 4 (1935) 467, pro pl. extr.-or. —*C. majalis* var. *manshurica* Kom. Malyi opred. rast. Dal'novost. kraya [Small Key to Plants of Far East. Region] (1925) 153.

Described from Japan. Type in Leiden (L) ?

Coniferous and deciduous forests, forest meadows and fringes.

IA. Mongolia: *East. Mong.* (Muni-Ula, nor. slope, in forest, common and abundant, July 5, 1871—Przew.).

General distribution: Far East, Nor. Mong. (Fore Hing.), China (Dunbei, North), Korean peninsula, Japan, Nor. Amer.

Note. Distinguished from common lily of the valley—*C. majalis* L.—in well-developed bracts, equal or longer than pedicel, and much larger broad cyathiform perianth.

Paris L.
Sp. pl. (1753) 367.

P. quadrifolia L. Sp. pl. (1753) 367; Kryl. Fl. Zap. Sib. 3 (1929) 657; Knorr. in Fl. SSSR, 4 (1935) 469, p.p.; Grub. Konsp. fl. MNR (1955) 96; Fl. Kazakhst. 2 (1958) 230.

Described from Europe. Type in London (Linn.).

Coniferous and deciduous forests, in birch groves, forest fringes, coastal scrubs.

Found in contiguous areas of Nor. Mongolia.

General distribution: Europe, Mediterr., West. and East. Sib., Nor. Mong. (Hent., Mong.-Daur.).

19. Smilax L.
Sp. pl. (1753) 1028.

1. S. stans Maxim. in Bull. Ac. Sci. St.-Pétersb. 17 (1872) 170; Forbes and Hemsley, Index Fl. Sin. 3 (1903) 101; Walker in Contribs. U.S. Nat. Herb. 28 (1941) 603.

Described from Japan. Type in Leningrad.

Among scrubs, in forests and forest fringes.

IIIA. Qinghai: *Nanshan* ("Shui Mo Kou, near Lien Cheng in woods, No. 346, June 1923, R.C. Ching"—Walker, l.c.).

General distribution: China (North-west, South-west), Japan.

Note. This is perhaps only a variety of *S. vaginata* Decne.

Family 25. DIOSCOREACEAE R. Br.

1. Dioscorea L.
Sp. pl. (1753) 1032; Knuth in Engler's Pflanzenr. IV, 43 (1924) 45; Prain and Burkill in Ann. Bot. Gard. Calcutta, 14, 1–2 (1936–1938) [Genus *Dioscorea* in the East].

1. D. nipponica Makino, Ill. Fl. Jap. 1 (1891) 2, tab. 45; Knuth in Engler's Pflanzenr. IV, 43 (1924) 314; Prain and Burkill in Ann. Bot. Gard. Calcutta, 14, 1 (1936) 61; Kitag. Lin. Fl. Mansh. (1939) 146. —*D. quinqueloba* auct. non Thunb.; Bunge, Enum. pl. China bor. (1832) 64; Franch. Pl. David. 1

(1884) 300; Forbes and Hemsley, Index Fl. Sin. 3 (1903) 92; Walker in Contribs. U.S. Nat. Herb. 28 (1941) 604. —*D. polystachya* auct. non Turcz.: Gontscharov in Fl. SSSR, 4 (1935) 497. —*D. giraldii* Knuth, l.c. 315. —**Ic.:** Makino, l.c.; Engler, Pflanzenr. IV, 43, fig. 60; Prain and Burkill, l.c. 2, pl. 24.

Described from Japan. Type in Tokyo.

Sparse forests and fringes, scrubs along montane slopes.

IA. Mongolia: *East. Mong.* (Muni-Ula, June 1871—Przew.; "Tatsing-Shan [Datsinshan'], Halaching-Kow, 1200 m, No. 2684; near Watachao, 1700 m, No. 2999—W.I. Hsia"—Prain and Burkill, l.c.).

IIIA. Qinghai: *Nanshan* ("Shui Mo Kou near Liencheng on a bushy slope, No. 357, 1923, Ching"—Walker, l.c.).

General distribution: Far East (south.), China (Dunbei, North, North-west, Cent., East, South-west), Korean peninsula, Japan.

Note. Var. *rosthornii* (Diels) Prain et Burk., more often found in our region, differs incharacteristic from typical one in leaves, greyish above (not bright green) and diffusely pubescent not only beneath but also above.

Family 26. AMARYLLIDACEAE Jaume

1. Ixiolirion Herb.
Appendix (1821) 37.

1. **I. tataricum** (Pall.) Herb. l.c.; Rgl. in Acta Horti Petrop. 6, 2 (1880) 492; Kryl. Fl. Zap. Sib. 3 (1929) 659; Gorschkova in Fl. SSSR, 4 (1935) 490; Fl. Kirgiz. 3 (1951) 123; Fl. Kazakhst. 2 (1958) 231; Grub. in Bot. mat. (Leningrad) 19 (1959) 536; Fl. Tadzh. 2 (1963) 366. —*I. kolpakowskianum* Rgl. l.c. 494. —*Amaryllis ? tatarica* Pall. Reise, 3 (1776) 727. —*Kolpakowskia ixioliroides* Rgl. in Acta Horti Petrop. 5, 2 (1878) 634. —**Ic.:** Pall, l.c. tab. D, fig. 1; Herb. Amaryllid. (1827) tab. 19; Fl. SSSR, 4, Plate XXX, fig. 4; Fl. Kazakhst. 2, Plate XXI, fig. 3; Fl. Tadzh. 2, Plate LXIV, figs. 2–3.

Described from West. Kazakhstan (Indersk lake). Type in Leningrad.

Steppe slopes and desert-steppe trails of mountains, 800–1500 m alt.

IIA. Junggar: *Tien Shan* (East. and Nor.); *Jung. Gobi* (shoulders of Urumchi-San'tai highway in San'tai region, semidesert steppe, 800 m, May 20, 1952, Mois.—far east. reports); *Zaisan* (Ch. Irtysh valley, Mai-Kapchagai area, June 1, 1903—Gr.-Grzh.); *Dzhark.*

General distribution: Aralo-Casp., Fore Balkh., Jung.-Tarb., Nor. Tien Shan; Mid. Asia.

Family 27. IRIDACEAE Juss.

1. Plant acaulous but with high hollow sheaths surrounding long perianth tube, with bulb covered with scarious-fibrous tunic with parallel fibres. Leaves filiform-linear. Perianth lobes similar; style

branched into filiform lobes ....
.... 1. **Crocus** L. (*C. alatavicus* Semen. et Rgl.).
+ Plant generally with distinct stem, rhizomatous or with bulb,
   covered with reticulate-fibrous tunic. Lobes of inner and outer
   perianth whorls different; lobes of tripartite style petaloid,
   protecting anthers .... 2. **Iris** L.

1. Crocus L.
Sp. pl. (1753) 36.

1. **C. alatavicus** Semen. et Rgl. in Bull. Soc. natur. Moscou, 2 (1868) 434;
Rgl. in Acta Horti Petrop. 6, 2 (1880) 498; B. Fedtsch. in Fl. SSSR, 4 (1935)
505; Fl. Kirgiz. 3 (1951) 124; Fl. Kazakhst. 2 (1958) 233; Fl. Tadzh. 2 (1963)
371. —Ic.: Fl. Kazakhst. 2, Plate XXI, fig. 1.
  Described from Kazakhstan (Jung. Alatau). Type in Leningrad.
  Hill meadows and steppes.

IIA. Junggar: *Tien Shan* (nor.-west.: Borborogusun picket, 1200 m, March 2; from
Almaty gorge to Kokkamyr, 1200–2100 m, April 28—1878; Louzogun, 760 m, March
8; Talki, 900 m, March 9; Pilyuchi, 900–1500 m, April 22; upland at sources of
Dzhirgalan and Pilyuchi, 1800 m, April 24—1879, A. Reg.; Kungess Tal, May 1–5,
1908—Merzb.).
  General distribution: Jung. Alatau, Nor. Tien Shan; Mid. Asia (West. Tien Shan,
Syr Darya).

2. Iris L.
Sp. pl. (1753) 38.

1. Plant bulbous, covered with reticulate-fibrous tunic. Leaves
    narrowly linear or filiform .... 21. **I. kolpakowskiana** Rgl.
+   Plant rhizomatous .... 2.
2. Stem in upper part dichotomously branched. Flowers in racemose
    inflorescence, small. Leaves broad, acinaciform, arranged in same
    plane. Seeds winged .... 20. **I. dichotoma** Pall.
+   Stem unbranched. Flowers single or 2–5 together. Seed wingless.
    .... 3.
3. Leaves very long and gyrose, often longitudinally convoluted,
    linear-filiform, 1–1.5 mm broad. Plant acaulous: spathe and
    flower barely exserted from compact mat ...... 9. **I. tenuifolia** Pall.
+   Leaves flat, neither convoluted nor gyrose .... 4.
4. Rhizome thick, 10–15 mm in diam., creeping, glabrous, with very
    short remnants of leaf sheaths only at tip. Leaves 10–16 mm
    broad, erect, ensiform or acinaciform, greyish or glaucous .... 5.

+ Rhizome slender, long, funiform or very short, densely covered with long fibrous remnants of leaf sheaths. Leaves narrow, up to 8 mm broad, green ....6.

5. Leaves erect, ensiform, greyish-green, stiff. Plant large, 50–100 cm tall. Flowers light yellow. Capsule hexagonal, with alternate narrow and broad faces ....**2. I. halophila** Pall.

+ Leaves acinaciform, grey. Plant 5–20 cm tall. Flowers blue. Capsule strictly hexagonal .... 17. **I. scariosa** Willd.

6. Plant forming more or less compact mat; rhizome short, branched, many-headed ....11.

+ Plant not forming mat; rhizome creeping or ascending, stems single ....7.

7. Stems compact. Leaves graminaceous, slender, with clearly projecting oblong veins, acuminulate and narrow toward base. Rhizome funiform, creeping ....8.

+ Stems hollow. Leaves linear, compact, with poorly visible veins. Rhizome ascending, fairly thick ....10.

8. Stems nearly as tall as radical leaves. 20–60 cm tall. Inflorescence 2–4-flowered. Capsule fusiform, beaked .... 5. **I. maximowiczii** Grub.

+ Stem much shorter than radical leaves, 3–20 cm tall, with solitary flower in bilobed spathe at tip. Capsule globose-ovoid, without beak...9.

9. Spathe lobes acute, scarious, slender, light green, withering in fruit. Plant up to 50 cm tall, with 10–20 cm tall stem, rarely small (10–20 cm) with 3–7 cm tall stem (var. *nana* Maxim.) .... 7. **I. ruthenica** Ker-Gawl.

+ Spathe lobes obtuse or shortly acuminate, compact, membranous, yellowish with reddish tip, later subpergamentaceous, persistent in fruit. Plant invariably small, rarely taller than 20 cm. .... 10. **I. uniflora** Pall.

10. Stem longer than leaves, with 2–3 terminal flowers clustered in 2–3-lobed spathe; blades of outer perianth lobes gradually narrow into claw, obovate, dark blue, same colour as inner ones. Capsule length twice thickness .... 8. **I. sibirica** L.

+ Stem not longer than leaves, with 2–3 terminal flowers and sometimes with additional solitary undeveloped flower below; blades of outer perianth lobes narrow abruptly into claw, suborbicular, bright violet, inner ones dark blue. Capsule length thrice thickness .... **I. orientalis** Thunb.

11 (6). Plant 20–50 cm tall, forming large, very compact mat with mass of coarse-fibred, rust- or reddish-brown sheath remnants of dead

leaves 5–12 cm long. Leaves linear, stiff. Roots slender, filiform. Lobes of spathe coarsely scarious, 4–12 cm long; outer perianth lobes glabrous .... 12.

+ Plant 6–20 cm tall; mat small, low. Sheath remnants of dead leaves finely fibrous or scarious, colourless, light straw-coloured or grey, short. Leaves not stiff. Roots contractile, funiform, often thickened. Lobes of spathe scarious, up to 4 cm long; outer perianth lobes with longitudinal band of hairs (bearded) on upper surface .... 16.

12. Lobes of highly inflated spathe reticulately veined due to sharply projecting transverse anastomoses between longitudinal veins ... . 11. **I. ventricosa** Pall.

+ Lobes of spathe with only distinct longitudinal veins, not reticulately veined .... 13.

13. Spathe 8–15 cm long, usually longer than stem. Leaves narrow, 2–4 mm broad. Perianth tube longer than limb; pedicels very short .... 14.

+ Spathe 4–8 cm long, usually much shorter than stem. Leaves 4–8 mm broad. Perianth tube many times shorter than limb; pedicel long .... 15.

14. Spathe highly inflated; perianth tube as long as limb or slightly longer; pedicel much longer than ovary .... 1. **I. bungei** Maxim.

+ Spathe not inflated, narrow; perianth tube 2–4 times longer than limb, up to 15 cm long; pedicel not longer than ovary .... 4. **I. loczii** Kanitz.

15. Perianth lobes bluish-violet, rarely white, narrow, outer broader than inner and obtuse or rounded at tip; capsule oval-cylindrical, its length 2–3 times its thickness .... 3. **I. lactea** Pall.

+ Perianth lobes blue, nearly equally broad, acute; capsule fusiform-cylindrical, its length 3–6 times its thickness .... 6. **I. oxypetala** Bge.

16 (11). Spathe 3-lobed, 2- or sometimes 3-flowered .... 17.

+ Spathe 2-lobed and 1-flowered .... 19.

17. Leaves ensiform, up to 8 mm broad. Flowers yellow .... 18.

+ Leaves narrow, linear, up to 3 mm broad. Flowers blue .... 15. **I. pandurata** Maxim.

18. Spathe inflated; its lobes obtuse, with distinct oblong, interrupted veins, joined with distinct anastomoses, violet along margin .... 12. **I. bloudowii** Ledeb.

+ Spathe not inflated; its lobes acute, with poorly visible veins without anastomoses, not coloured along margin .... 13. **I. flavissima** Pall.

19. Rhizome and roots slender. Leaves slender, linear, graminaceous, usually shorter than stem. Tube of violet perianth less than 1/2 length of ovary .... 14. **I. goniocarpa** Baker.

+ Rhizome thick, roots funiform, contractile. Leaves compact, narrowly, longer than stem, or plant acaulous. Perianth tubes longer than ovary .... 20.

20. Fibrous remains of dead leaves in mat crispate-sinuate; leaves obtuse or shortly acuminate. Flowers violet or pale yellow, fragrant .... 18. **I. thoroldii** Baker.

+ Fibrous remains of dead leaves erect; leaves acuminulate or acute. Flowers odourless .... 21.

21. Root dark grey, very thick and long. Flowers bright violet, on very short stem; perianth tube distinctly shorter than limb .... 19. **I. tigridia** Bunge.

+ Root light coloured, very slender. Flowers light yellow, sessile, with long tube, longer (sometimes twice) than limb .... 16. **I. potaninii** Maxim.

Section 1. Apogon Baker

1. **I. bungei** Maxim. in Bull. Ac. Sci. St.-Pétersb. 26 (1880) 509; Franch. Pl. David. 1 (1884) 296; Forbes and Hemsley, Index Fl. Sin. 3 (1903) 80; Dykes, Gen. Iris (1913) 34; Pavl. in Byull. Mosk. obshch. ispyt. prir., Biol. 38, 1–2 (1929) 41; Grub. Konsp. fl. MNR (1955) 96; Hanelt and Davazamè in Feddes repert. 70, 1–3 (1965) 20. —*I. ventricosa* auct. non Pall.: Maxim. Prim. fl. amur. (1859) 485 (Index fl. mong.); ? Walker in Contribs. U.S. Nat. Herb. 28 (1941) 604.

Described from East. Mongolia. Type in Leningrad. Plate VI, fig. 2; map 5.

Dry and common cattail and feather-grass desert steppes and loamy sand deserts, on sandy and gravelly soil, on rubble and rocky slopes, rocks, floors and flanks of gorges.

IA. **Mongolia:** *Cen. Khalkha* (far nor., reports: 8–10 km south of Sergulen somon along road to Dalan-Dzadagad, June 4, 1949—Yun.); *East. Mong.* (Shabartu plant [nor. of Kalgan] May 21, 1831 [I. Kuznetsov], typus!); *Val. Lakes* (far west. reports: between Baidarik river and Chintologoi cliff, on steppe, June 22, 1894—Klem.); *Gobi-Alt.* (Dundu-Saikhan mountain range, south. slope, midbelt, on rocky slope, July 13, 1909—Czet.); *East. Gobi* (including desert between Suma-Khada and Nanshan mountain ranges [at Khoir-Obo between Dabkhyr river and Dabasun-Nor], June 17, 1871; Huang He valley, solonetz-sandy desert [in Burkhan-Bulak region], in thickets, common, May 11, 1872, Przew., paratypes!); *Alash Gobi* (Galbyn-Gobi, Khara-Morite area, near well, May 15, 1909—Czet.; road to Kuku-Nor from Alashan, Tarbagatai area, around sand-dunes, June 7, 1908—Czet.; 20 km nor. of Taodaokhe settlement, sandy-pebble plain with *Reaumuria*-scrub vegetation, June 15, 1958—Petr.; "Wang Ieh Fu [Ting Yüan Ying]; Nan Ssu Kou. Large compact clusters along dry sandy roadsides and on moist rich farmland—R.C. Ching"—Walker, l.c.).

General distribution: China (? North: Beijing vicinity, 1840–1850—Tatarinow; North-west: nor. part of Gansu province, along road to Ningan'pu from Sykuz, April 20, 1909—Demidenko).

Note. The herbarium specimen of this species from Tatarinow's collection bears only the standard printed label (Fl. Pekin, Dr. Tatarinow, acc. 1856), without date or locality. A.A. Tatarinow probably collected it in 1840 in East. Mongolia on the way to Beijing from Kyakhta or in 1850 on the way back from Beijing to Kyakhta and the wrong label was attached.

2. **I. halophila** Pall. Reise, 2 (1773) and 3 (1776) 713; Kryl. Fl. Zap. Sib. 3 (1929) 665; B. Fedtsch. in Fl. SSSR, 4 (1935) 526; Fl. Kazakhst. 2 (1958) 242; Grub. in Novosti sist. vyssh. rast. 6 (1970) 35. —*I. sogdiana* Bge. Rel. Lehmann. (1851) 507; B. Fedtsch. in Fl. SSSR 4 (1935) 526; Fl. Kirgiz. 9 (1951) 131; Fl. Kazakhst. 2 (1958) 242; Fl. Tadzh. 2 (1963) 376. —*I. gueldenstaedtiana* Lepech. in Acta Ac. Sci. Petrop V, 1 (1784) 292; Maxim. in Bull. Ac. Sci. St.-Pétersb. 26 (1880) 518, p.p., incl. var. *sogdiana* Maxim.; id. in Acta Horti Petrop. 6, 2 (1880) 496; Boiss. Fl. or. 5 (1884) 129. —*I. spuria* var. *halophila* Dykes, Gen. Iris (1913) 62. —**Ic.**: Pall. l.c. 3, tab. B, fig. 2; Acta Ac. Sci. Petrop. V, 1, tab. 8; Fl. Tadzh. 2, Plate LXVI, figs. 1–2.

Described from West. Siberia (Kulund steppe). Type in London (BM).

Solonetz coastal and steppe meadows, chee grass thickets on solonetz banks of rivers, lakes and irrigation ditches, among steppe scrubs, from desert to subalpine belt.

IB. **Kashgar**: *Nor.* (Uchturfan, Kukurtuk gorge, May 27, 1908—Divn.; Baisk basin, left bank portion of debris cone of Muzart river near Bazar village, fallow wormwood zone with iris groves, Sept. 3, 1958—Yun; Baisk basin 1 km south of Daban'chen pass, 1440 m, desert, No. 8180, Sept. 3, 1958—Lee and Chu); *East* (west of Khami, 850 m, near water, No. 5405, May 22, 1958—Lee and Chu).

IIA. **Junggar**: *Cis-Alt.* (near Kran river [7–10 km before Shara-Sume], Aug. 29, 1876—Pot.; same locality, 15–20 km nor.-west of Shara-Sume, scrub steppe, July 7, 1959—Yun.); *Tarb.* (Khobuk river valley, 10–12 km south of Khobuk settlement, among thickets, June 24, 1957—Yun.); *Jung. Alatau* (near Toli town, meadow steppe, No. 2487, Aug. 4, 1957—Kuan; Argaty river valley between Borotala and Arasan, poplar forest of floodplain, Aug. 17, 1957—Yun.); *Tien Shan* (Nor. and East.); *Jung. Gobi* (south., east up to Guchen; east., Oshigiin-Usu area 15–20 km north of Baitak-Bogdo mountain range, solonchak chee grass thicket, July 30, 1941—Yun., far east. report); *Dzhark.* (Ili river-bank west of Kul'dzha, May 1877—A. Reg.).

General distribution: Aralo-Casp., Fore Balkh., Jung.-Tarb., Nor. and Cent. Tien Shan; Europe (Black Sea region and Fore Caucasus), Mid. Asia (hill.), West. Sib. (south).

3. **I. lactea** Pall. Reise, 3 (1776) 713; Grub. in Novosti sist. vyssh. rast. 6 (1970) 32.—*I. biglumis* Vahl, Enum. pl. 2 (1806) 149; Kitag. Lin. Fl. Mansh.

(1939) 146; Poljakov in Bot. mat. (Leningrad) 12 (1950) 90; Grub. Konsp. fl.
MNR (1955) 96; Hanelt and Davazamè in Feddes Repert. 70, 1–3 (1965)
20. —*I. pallasii* Fisch. in Trevir. Ind. Sem. Horti Vratisl. (1821); Poljakov in
Bot. mat. (Leningrad) 12 (1950) 89; Grub. Konsp. fl. MNR (1955) 97. —*I.
haematophylla* Link, Enum. pl. horti Berolin. 1 (1821) 60, non Fisch. (1823);
Fl. Kazakhst. 2 (1958) 238. —*I. iliensis* Poljak. in Bot. mat. (Leningrad) 12
(1950) 88; Fl. Kazakhst. 2 (1958) 236. —*I.* an *spuria*? Pall. Reise, 3 (1776)
713. —*I. ensata* auct. non Thunb.: Maxim. in Bull. Ac. Sci. St.-Pétersb. 26
(1880) 512, p.p.; id. in Acta Horti Petrop. 6, 2 (1880) 496, p. max. p.; Franch.
Pl. David. 1 (1884) 297, p.p.; quoad pl. gehol.; ?Kanitz, A novenytani...(1891)
58; Diels in Futterer, Durch Asien (1903) 8; Forbes and Hemsley, Index Fl.
Sin. 3 (1903) 81, p.p.; Palibin in Tr. Troitskosavsko-Kyakht. otdelen. RGO
[Russian Geographic Society], 7, 3 (1904) 41; Dykes, Gen. Iris (1913) 85,
p.p.; Danguy in Bull. Mus. nat. hist. natur. 17 (1911) 557 and 20 (1914)
137; Kryl. Fl. Zap. Sib. 3 (1929) 662; Pavl. in Byull. Mosk. obshch. ispyt.
prir., Biol. 38 (1929) 41; B. Fedtsch. in Fl. SSSR, 4 (1935) 518, p. max. p.;
Walker in Contribs. U.S. Nat. Herb. 28 (1941) 604; ?Chen and Chou, Rast.
pokrov r. Sulekhe [Vegetation Cover of Sulekhe River] (1957) 92. —**Ic.:** Pall.
l.c. tab. C, fig. 1 (sub *I.* an *spuria* L.?); Reichb. Ic. pl. crit. V, tab. 479 (sub *I.
pallasii*); Fl. Kazakhst. 2, Plate XXIII, fig. 1 (sub *I. haematophyllae*), fig. 3 (sub
I. iliensis).

Described from Dauria (Tare lake vicinity). Type in London (BM)?

Solonetz banks of rivers, lakes, springs, sandy floors of gorges, solonetz
coastal and steppe meadows, floors of valleys, along edges of solonchaks,
saline hummocky marshes, chee grass thickets on solonetz rather thin
sand, up to 3000 m alt.; often in thickets.

IA. **Mongolia:** *Khobd., Mong. Alt., Cen. Khalkha, East. Mong., Depr. Lakes, Val.
Lakes, Gobi-Alt., East. Gobi* (valley of Huang He river, in desert, rare, May 14, 1872—
Przew.); *Alash. Gobi* (solonetz sand bank of Dzharatai lake, June 3, 1872—Przew.;
Kharmykten-Khuduk area, sand bed, May 9; Galbyn-Gobi, Khara-Morite area, near
well on moist soil, May 15; Edzin-Gol river, Dzhargalante area, along valley, July 17—
1909, Czet.; Valley of Edzin-Gol river [near Ontsin-Gol upper course], Bukhankhub
area, on loam, June 10, 1926—Glag.; "Wangye-fu, large pure stands along roadsides,
in either wet or dry places, 1923, R.C. Ching"—Walker, l.c.); *Khesi* (between Gaotai
and Fuetin towns, on solonchak, June 12, 1886—Pot.; "Kantschou-fu [Chzhan'e], May
28, 1879"—Kanitz, l.c.; "Oase von Astün, zwischen Lager XXI und Sutschou"—Diels,
l.c.; "Sulekhe river valley"—Chen and Chou, l.c.).

IC. **Qaidam:** *plains* (south. fringe of Barun-Tszasaka hyrma, 2600 m, May 5, May
20 and June 3; nor. slope of Burkhan-Budda mountain range, Nomogun-Gol gorge,
2900 m, May 17, 1900—Lad.).

IIA. **Junggar:** *Tien Shan* (south. bank of Sairam lake, July 20, 1877—A. Reg.); *Jung.
Gobi* (nor.: Barbagai, state farm of 28th regiment, 550 m, Sept. 7, 1956—Ching; nor.-
west.: Khobuk-Sair, in humid gorge, No. 3344, Sept. 22, 1957—Ching; west.: Ebi-Nur
vicinity, in desert, No. 4537, Aug. 18, 1957—Kuan); *Dzhark.* (Ili river-bank south of
Kul'dzha, April 30, 1877—A. Reg.).

IIIA. Qinghai: *Nanshan* (South Kukunor mountain range, 3650 m, April 29, 1884—Pot.; "Liufuyai—1923, R.C. Ching"—Walker, l.c.); Amdo (along Baga-Gorgi river, 2750 m, common, May 23; in alpine belt of Dzhakhar-Dzhargyn mountain range, June 24—1880, Przew.; environs of Kazhir village on Karyn river, May 5, 1885—Pot.).

General distribution: Fore Balkh., ? Jung.-Tarb.; West. Sib. (south-east.), East. Sib. (south-east.), Nor. Mong. (Hent., Hang., Fore Hing., Mong.-Daur.), China (Dunbei: nor.-west., South-west nor. Gansu).

4. **I. loczyi** Kanitz, A. novenytani...(1891) 58; id. in Die Resultate der botanische Sammlungen, 2 (1898) 733; Forbes and Hemsley, Index Fl. Sin. 3 (1903) 83; Ikonnikov, Opred. rast. Pamira [Key to Plants of Pamir] (1963) 89. —*I. tenuifolia* Pall. var. *tianschanica* Maxim. in Bull. Ac. Sci. St.-Pétersb. 26 (1880) 512; id. in Acta Horti Petrop. 6, 2 (1880) 495; Dykes, Gen. Iris (1913) 32. —*I. tianschanica* (Maxim.) Vved. in Fl. Turkmenii, 1, 2 (1932) 325; B. Fedtsch. in Fl. SSSR, 4 (1935) 515; Fl. Kirgiz. 3 (1951) 130; Fl. Kazakhst. 2 (1958) 237; Fl. Tadzh. 2 (1963) 375. —*I. tenuifolia* auct. non Pall.: Diels in Filchner, Wissensch. Ergebnisse (1908) 249; ? Danguy in Bull. Mus. nat. hist. natur. 17 (1911) 449; Persson in Bot. notiser (1938) 277; Walker in Contribs. U.S. Nat. Herb. 28 (1941) 604, p.p., quoad pl. nanschan. —**Ic.:** Kanitz, l.c. (1891) tab. VI, fig. II, 5; Fl. SSSR, 4, Plate XXXI, fig. 12, Plate XXXII, fig. 5.

Described from Qinghai (Nanshan). Type in Vienna (W).

In montane, more often wormwood-forb steppes on loessial, clayey and clayey-rubble slopes, coastal sandy-pebble shoals and rock crevices, from 1000 (in Nor. Tien Shan) and 2000 to 4200 m, often in groves.

IB. Kashgar: *Nor.* (nor.-west. slope of Sogdyntau mountain range 10–12 km southeast of Akchii water monitoring station west of Aksu, 2600 m, wormwood-forb steppe along gullies, Sept. 19. 1958—Yun.; same locality, Akchii, No. 8433, Sept. 19, 1958—Lee and Chu (A.R. Lee (1959)); *West.* (upper Tiznaf river, Akkëz crossing along road to Tibet from Jarkend, 3400 m, steppe along loessial ridges, June 5, 1959—Yun.; same locality, June 5, 1959—Lee and Chu); nor. slope of Kingtau mountain range 1 km north of Koshkulak settlement, wormwood-Koeleria steppe, June 10, 1959—Yun.; same locality, on nor. slope, No. 00218, June 10, 1959—Lee and Chu); along Kashgar-Torugart road, on mountain top, 3500 m, No. 9746, June 20, 1959—Lee and Chu); *South., East.* (Turfan district, 8 km nor. of San'shan'kou, 2450 m, steppe on exposed slope, No. 5643, June 15; along road to Shikhotsze from San'shan'kou, steppe on shady slope, No. 5657, June 16; along Turfan-Kalangou road, in shade, No. 5826, June 24, 1958—Lee and Chu).

IC. Qaidam: *hill.* (south. slope of South-Kukunor mountain range, Serin area, 3350–3650 m, June 4, 1895—Rob.).

IIA. Junggar: *Tien Shan* (along Sochzhan road, June 13, 1877—Pot. [far east. report]; M. Yuldus upland, 2300 m, on moist loam, June 14, 1877—Przew.; Almaty valley around Kul'dzha, 1000–1800 m, April 25; Talki river near Kul'dzha, 1000–1200 m, May—1878, A. Reg.; Dzhanart river valley, June 14–17, 1903—Merzb.; nor. slope of Bogdo-Ula along road to montane lake from Fukan, east. steppe slope near lower boundary of spruce forests [wormwood-feather-grass steppe], April 26, 1959—Yun.; B. Yuldus basin 10–12 km nor.-west of Bain-Bulak settlement along road to Dachit crossing, steppe along south. slope of Narat mountain range, Aug. 8, 1958—Yun.).

IIIA. **Qinghai:** *Nanshan* (many reports including: "In m. Nanszan collinis circa Cziakouye, June 10, 1879"—Kanitz, l.c., typus!); *Amdo.*

IIIB. **Tibet:** *Weitzan, South.* ("Khambajong, 1903—Younghusband"—Dykes, l.c.).

IIIC. **Pamir** (Taret pass, June 14, 1909—Divn.; crossing from Arpalyk river gorge to Kizil-Bazar, 3500 m, July 13, 1941; Yazag and Balung river sources, 4000 m, June 16, 1942; Pil'nen river, 3000–4000 m, June 30, 1942—Serp.).

General distribution: Jung.-Tarb., Nor. and Cent. Tien Shan, East. Pamir; Mid. Asia (West. Tien Shan, Pam.-Alay).

5. **I. maximowiczii** Grub. sp. nova. —*I. songarica* var.? *gracilis* Maxim. in Bull. Ac. Sci. St.-Pétersb. 26 (1880) 510; Forbes and Hemsley, Index Fl. Sin. 3 (1903) 84. —*I. songarica* auct. non Schrenk: Dykes, Gen. Iris (1913) 71.

I. soongoricae Schrenk proxima, sed foliis tenerioribus, 3–6 mm latis, radicalibus cauli aequilongis vel paulo superantibus, spathis angustioribus, perigonii tubo limbo triplo vel quadriplo breviore ovarium aequante, pedicello capsulae aequilongo vel eam duplo superante bene distincta.

Typus: Tsinghai, Nan-Schan, jugum a fluvio Tetung boream versus in vallibus rivulorum alpinorum, No. 129, July 4, 1872, Przewalski; in Herb. Inst. Bot. Acad. Sci. URSS (Leningrad) conservatur.

Plate VI, fig. 1; map 5.

Meadow valleys of rivers and among scrubs in alpine mountain belt.

IIIA. **Qinghai:** *Nanshan* (North Tetung mountain range in alpine valleys of rivers, uncommon, No. 129, July 4, 1872, Przew., typus!; mountain range between Nanshan and Donkyr on Raka-Gol river, in alpine scrubs, occasional, July 22, 1880—Przew.; Matisy temple 70 km south-east of Chzhan'e town, Tsilin'shan' mountain range, 2600 m, scrubs and small meadows in mountain valley, July 12, 1958—Petr.).

General distribution: China (South-west—nor. Sichuan: along Nereku river, July 26 and on rocks along Atu-Lunva river, Aug. 9—1885, Pot.).

Note. This species was noticed by K.I. Maximowicz (l.c.) but he desisted from describing it as a species because of lack of adequate and complete herbarium material.

I. orientalis Thunb. in Trans. Linn. Soc. London, 2 (1794) 328; Maxim. in Bull. Ac. Sci. St.-Pétersb. 26 (1880) 519; Dykes, Gen. Iris (1913) 23; B. Fedtsch. in Fl. SSSR, 4 (1935) 520. —*I. sibirica* var. *orientalis* Baker, Handb. Irid. (1892) 9; Forbes and Hemsley, Index Fl. Sin. 3 (1903) 84. —*I. nertschinskia* Lodd. Bot. Cab. (1833) tab. 1845; Kitag. Lin. Fl. Mansh. (1939) 148. —**Ic.:** Gartenfl. (1813) tab. 1604; Lodd. l.c.; Dykes, l.c. tab. 1, b; Fl. SSSR, 4, Plate XXI, fig. 14.

Described from Japan. Type in Upsala (UPS).

Humid and marshy meadows and marshes, along river-banks.

Reported from adjoining Dunbei region (near Yakshi railway station, marsh, June 13, 1902—Litw.). Its occurrence possible in **IA. Mongolia:** *East. Mong.*

General distribution: East. Sib. (Daur.), Far East (south.), China (Dunbei, ?North), Korean peninsula, Japan.

Note. This species is very closely related to *I. sibirica* L. and is difficult to distinguish from the latter.

6. I. oxypetala Bge. Enum. pl. China bor. (1832) 63, non C.A. Mey, 1834; Grub. in Novosti sist. vyssh. rast. 6 (1970) 33. —*I. fragrans* Lindl. in Bot. Reg. 26 (1840) tab. 1. —*I. moorcroftii* Wall. ex D. Don in Trans. Linn. Soc. London, 18 (1841) 315; Fl. Tadzh. 2 (1963) 378. —*I. pallasii β. chinensis* Fisch. ex Sims in Bot. Mag. 49 (1822) tab. 2331. —*I. ensata* var. *chinensis* Maxim. in Gartenfl. (1880) 161; id. in Bull. Ac. Sci. St.-Pétersb. 26 (1880) 514, excl. syn. *I. lactea* Pall. —*I. ensata* auct. non Thunb.: Maxim. in Acta Horti Petrop. 6, 2 (1880) 496, p. max. p.; Franch. Pl. David. 1 (1884) 297, p.p., excl. pl. gehol.; Forbes and Hemsley, Index Fl. Sin. 3 (1903) 81, p.p.; Danguy in Bull. Mus. nat. hist. natur. 17 (1911) 449; Dykes, Gen. Iris (1913) 85, p.p.; Pampanini, Fl. Carac. (1925) 89; Kashyap in Proc. 19 Indian Sci. Congr. (1932) 52; B. Fedtsch. in Fl. SSSR, 4 (1935) 518; p. min. p.; Persson in Bot. notiser (1938) 278; Poljakov in Bot. mat. (Leningrad) 12 (1950) 88; Vorob'ev, Opred. rast. Primor'ya i Priamur'ya [Key to Plants of Primor'e and Fore Amur] (1966) 128. —*I. pallasii* auct. non Fisch.: Kitag. Lin. Fl. Mansh. (1939) 148. —*I. iliensis* auct. non Pol.: Grub. in Bot. mat. (Leningrad) 19 (1959) 536. —**Ic.:** Bot. Mag. 49, tab. 2331; Bot. Reg. 26, tab. 1; Gartenfl. 29, tab. 1011.

Described from Nor. China (Beijing vicinity). Type in Leningrad.

In meadows (more often, solonetzic) along river-banks and valley floors, on solonetz banks of rivers, lakes and springs, along fringes of solonchak swamps, solonetz sand, chee grass thickets as well as along canals, irrigation ditches and borders of irrigated fields, up to 3500 m (?), often in thickets.

IA. Mongolia: *East. Mong.* (south.: plain in vicinity of Kuku-Khoto, July 6, 1884—Pot.); Alash. Gobi (Fuinchzha around Bayan-Khoto town, in short-grass meadows, common, July 6, 1957—Kabanov; around Bayan-Khoto, Otirloso area, 120 km nor. of Tsilintai settlement, sand-covered solonchak, July 8, 1957—Petr.; Alashan mountains, solonchak Bain-Bulak swamp, June 18; same locality, west. slope in middle mountain, belt, humid mountain valley, July 3—1873, Przew.; same locality, in grasslands, April 25, 1908—Czet.); *Ordos* (around Linchzhu village, Oct. 6, 1884—Pot.); Khesi (nor. slope of mountain range not far from Dangu town, humid grassy valley, common and in beds, June 2, 1872—Przew.; along Lonsyr river around Liyuan'in village, along valley floor, June 2, 1886—Pot.; Sachzhou oasis, 1100 m, along solonetz moist lake-bank, thickets, June 22, 1894—Rob.).

IB. Kashgar: *Nor.* ("Province de Koutchar [Kucha], Goum-Toura, April 23, 1907, Vaillant"—Danguy, l.c.; around Uchturfan, Kichakaral, bank of irrigation ditch, May 10, 1908—Divn.; west. part of Baisk basin, left half of debris cone of Muzart river near Bazar village, Sept. 3, 1958—Yun.; Bai, 1 km south of Daban'chen, 1440 m, desert, No. 8181, Sept. 3, 1958—Lee and Chu); *West., South.* (Keriyadar'ya near Lyushi estuary, 2900 m, in humid gorge, common, June 30, 1885—Przew.; nor. foothills of Kunlun, 1500 m, along irrigation ditches in gardens, ? 1889—Rob.; 20 km west of Keriya town along road to Khotan, solonchak-like chee grass thickets, May 16, 1959—Yun.; same locality, semistable sand, No. 121, May 16, 1959—Lee and Chu); *East.* (in east up to Khami).

IIA. **Junggar:** *Tien Shan* (cent.: near Yagash-Gumbos, in steppe, No. 433, April 23; east of Yagash-Gumbos, in steppe, very common, No. 434, April end; between Kizyl-Gumbos and Uch, in steppe, No. 435, April 20–25—1903, Merzb.; Chzhaosu-Tekes, on slope, No. 3597, Aug. 16, 1957—Kuan; 30 km east of Bain-Bulak, intermontane basin, 2620 m, No. 6581, Aug. 11, 1958—Lee and Chu); *Jung. Gobi* (south.: from Sadab to Urumchi, 1200 m, Nov. 3, 1879—A. Reg.; near Santai on bank of irrigation ditch, may 25, 1952—Mois.; 7–8 km south-south-west of Urumchi, near Yanervo settlement, Urumchinki river valley, humid short-grass meadows in poplar bottom-land desert forest, May 31, 1957—Yun.; same locality, No. 541, May 31, 1957—Kuan; Fukan town vicinity, along field border, No. 4207, Sept. 17, 1957—Kuan); *Dzhark.* (Kul'dzha, 1876—Golike; same locality, April 27, 1877—A. Reg.; same locality, 1906—Muromskii, Khoyur-Sumun, south of Kul'dzha, May 27, 1877; near Talki river estuary around Suidun, June 7, 1878—A. Reg.).

IIIA. **Qinghai:** *Nanshan* (? South Tetung mountain range, lower and middle belts, in valleys, common and in clusters, May 26, 1873—Przew.; ? Gumansy temple, May 8, 1890—Gr.-Grzh.); *Amdo* (in Churmyn river gorge, on shoal, common, May 17, 1880—Przew.; Guidui town, 2200 m, May 7; valley of Lonchzhu river near Rtygri village, May 8—1885, Pot.).

IIIB. **Tibet:** ? *Weitzan, South.* ("Gyantze, 1904, Walton"—Dykes, l.c.; "Gyantze, 3950 m, July 1930"—Kashyap, l.c.).

General distribution: Mid. Asia (Syrdar., nor.-west. Pam.-Alay), Far East (south. Ussur.), China (east. Dunbei, North, North-west, South-west: nor. Yunnan), Himalayas (west., Kashmir), Korean peninsula.

7. **I. ruthenica** Ker-Gawl. in Bot. Mag. 28 (1808) tab. 1123; Maxim. in Bull. Ac. Sci. St.-Pétersb. 26 (1880) 516, incl. var. *brevituba* Maxim. and var. *nana* Maxim.; id. in Acta Horti Petrop. 6, 2 (1880) 496; Franch. Pl. David. 1 (1884) 297; Forbes and Hemsley, Index Fl. Sin. 3 (1903) 80; Dykes, Gen. Iris (1913) 52, p.p., excl. *I. unifloro* Pall.; Kryl. Fl. Zap. Sib. 3 (1929) 663; Pavl. in Byull. Mosk. obshch. ispyt. prir., Biol. 38 (1929) 41; B. Fedtsch. in Fl. SSSR, 4 (1935) 517; Kitag. Lin. Fl. Mansh. (1939) 148; Grub. Konsp. fl. MNR (1955) 97; Popov, Fl. Sr. Sib. 1 (1957) 205, excl. syn. and adnot.; Fl. Kazakhst. 2 (1958) 237; Grub. in Bot. mat. (Leningrad) 19 (1959) 536. —*I. brevituba* Vved. in Fl. Kirgiz. 3 (1951) 131; Fl. Kazakhst. 2 (1958) 238. —**Ic.:** Bot. Mag. 28, tab. 1123, 1393; Fl. Kazakhst. 2, Plate XXII, fig. 5.

Described from cultivated specimen. Type not preserved.

Subalpine and forest meadows, shaded steppe slopes and along steppe gullies, larch and spruce mountain forests, birch groves and forest fringes, 900–2900 m.s.m.

IIA. **Junggar:** *Tien Shan* (nor., cent. and east.).

General distribution: Fore Balkh. (east.), Jung.-Tarb., Nor. and Cent. Tien Shan; Mid. Asia (West. Tzien Shan), West. Sib., East. Sib., Far East (Ussur.), Nor. Mong. (Hent., Hang., Mong.-Daur.), China (Dunbei, North, South-West: Kam).

Note. There is no rationale for raising one of the two varieties—var. *brevituba* Maxim.—established by Maximowicz to the rank of species. The relative length of the perianth tube in *I. ruthenica* Ker-Gawl. is highly variable and no distinct demarcation is possible between long- and short-tubed forms. On the other hand, the distribution range of var. *brevituba* Maxim. is not confined exclusively to Tien Shan. Maximowicz (l.c.) himself indicated it for Altay and East. Siberia, apart from Tien Shan, while the treatment of specimens from Mongolian collections showed that this variety is found more often in Nor. Mongolia compared to the typical long-tubed variety. Thus, var. *brevituba* Maxim. enjoys no exclusive distribution range whatsoever and is found throughout the species distribution range. Along with it and the type variant, dwarfish var. *nana* Maxim. is also found in Nor. Mongolia, but far more rarely.

8. **I. sibirica** L. Sp. pl. (1753) 38; Maxim. in Bull. Ac. Sci. St.-Pétersb. 26 (1880) 519; Dykes, Gen. Iris (1913) 20; Danguy in Bull. Mus. nat. hist. natur. 20 (1914) 137; Kryl. Fl. Zap. Sib. 3 (1929) 666; B. Fedtsch. in Fl. SSSR, 4 (1935) 519; Grub. Konsp. fl. MNR (1955) 97; Fl. Kazakhst. 2 (1958) 240. —Ic.: Bot. Mag. 29, tab. 1163; Dykes, l.c. tab. 1, a; Fl. SSSR, 4, Plate XXXI, fig. 13; Fl. Kazakhst. 2, Plate XXII, fig. 3.

Described from Siberia and Europe. Type in London (Linn.).

Sand and foothills of sand-dunes with groundwater nearby, along river-banks and in coastal moist and marshy meadows, rare.

IA. **Mongolia:** *East. Mong.* ("Environs de Kailar, 750 m, sables, No. 1531, June 23, 1896—Chaff."—Danguy, l.c.; around Khailar town, near foot of sand hummocks, No. 633, June 10, 1951—S.H. Li et al.).

General distribution: Europe (Cent. and East.), Balk., Caucasus (excluding West. Transcaucasus), West. Sib., East. Sib. (Ang.-Sayan), Nor. Mong. (Hent., Mong.-Daur.).

9. **I. tenuifolia** Pall. Reise, 3 (1776) 714; Maxim. in Bull. Ac. Sci. St.-Pétersb. 26 (1880) 511; Franch. Pl. David. 1 (1884) 297; Forbes and Hemsley, Index Fl. Sin. 3 (1903) 85; Dykes, Gen. Iris (1913) 32, p.p., excl. var. *tianschanica* Maxim.; Danguy in Bull. Mus. nat. hist. natur. 20 (1914) 137; Kryl. Fl. Zap. Sib. 3 (1929) 662; Pavl. in Byull. Mosk. obshch. ispyt. prir., Biol. 38 (1929) 41; B. Fedtsch. in Fl. SSSR, 4 (1935) 515; Kitag. Lin. Fl. Mansh. (1939) 149; Walker in Contribs. U.S. Nat. Herb. 28 (1941) 604, p.p., quoad pl. mong.; Ching in Bull. Fan memor. Inst. Biol. (Bot.) 10, 5 (1941) 261; Grub. Konsp. fl. MNR (1955) 97; Fl. Kazakhst. 2 (1958) 236; Hanelt and Davazamè in Feddes repert. 70, 1–3 (1965) 20. —*I. regelii* Maxim. l.c. 526; id. in Acta Horti Petrop. 6, 2 (1880) 495. —Ic.: Pall. l.c. tab. C, fig. 2; Fl. SSSR, 4, Plate XXXII, fig. 4; Fl. Kazakhst. 2, Plate XXII, fig. 1.

Described from Dauria (Tare lake vicinity). Type in London (BM).

Sandy steppes, semistable and thin sand, steppe and desert-steppe rubble slopes of hummocks and hills.

IA. **Mongolia:** *Cent. Khalkha, East. Mong., Depr. Lakes* (13 km nor. of Khara-Nur [Khungui] lake, pea shrub-feather-grass steppe, Aug. 19, 1944—Yun.; "Steppe entre Kobdo et Ourga, Oct. 3, 1895, Chaff."—Danguy, l.c.); *Val. Lakes, Gobi-Alt.* (5 km west of Nomogon somon, sandy feather-grass steppe, April 27, 1941—Yun.); *East. Gobi, Alash. Gobi* (Alashan mountain range, Tsuburgan-Gol river, April 27, 1908—Czet.; "Holanschan, Tishuikou, 1923, R.C. Ching"—Walker, l.c.).

IIA. **Junggar:** *Tien Shan* (nor.: sources of Dzhirgalan river, 1800 m, April 24; Borborogusun river vicinity, 1800 m, June 15—1879, A. Reg.); *Jung. Gobi* (nor.: around Urungu river, April 24, 1879—Przew.; 30–35 km south-south-east of Shara-Sume along road to Shipati, wormwood-grass steppe, July 7, 1959—Yun.; nor.-west: Khobuk-Urungu interfluvine region, 35 km nor. of Shouege along road to Sulyugou, saxaul grove on thin sand, July 11, 1959—Yun.; east: Kukusyrkhe hills, on sand, May 21, 1879—Przew.; along Targyn-Gol river, Sept. 14, 1930—Bar.); *Dzhark.*

IIIA. **Qinghai:** *Nanshan* (Sanchuan' along Huang He river, April 7, 1885—Pot.).

General distribution: Aralo-Casp., Fore Balkh., Jung.-Tarb.; West. Sib. (far south-west), East. Sib. (Daur.), China (Dunbei, North, North-west: nor. Gansu).

10. **I. uniflora** Pall. ex Link in Spreng., Schrader and Link, Jahrb. 1, 3 (1820) 71; Maxim. in Bull. Ac. Sci. St.-Pétersb. 26 (1880) 517; Danguy in Bull. Mus. nat. hist. natur. 20 (1914) 137; B. Fedtsch. in Fl. SSSR, 4 (1935) 518; Kitag. Lin. Fl. Mansh. (1939) 149; Grub. Bot. zh. 56, 11 (1971) 1642. —*I. ruthenica* var. *dahurica* M. Pop. Fl. Sredn. Sib. 1 (1957) 205. —*I. ruthenica* auct. non Ker-Gawl.; Dykes, Gen. Iris (1913) 52, p.p. —**Ic.:** Fl. SSSR, 4, Plate XXXI, fig. 11.

Described from Dauria. Type in Berlin (B) ?

Dry meadows along slopes of mountains and river valleys, birch groves and forest meadows.

IA. **Mongolia:** *East. Mong.* (60 km south-east of Khamar-Daba settlement, *Leucopoa*-forb steppe on nor. slope, July 27, 1971—Z. Karamysheva and I. Safronova).

General distribution: East. Sib. (Daur.), Far East (continent.), Nor. Mong. (Fore Hing.), China (Dunbei nor.), Korean peninsula (far nor.).

Note. Notwithstanding M.G. Popov's (M. Pop. l.c.) affirmation, *I. uniflora* Pall. ex Link is not at all a mere synonym of *I. ruthenica* Ker-Gawl. Although not comparing this Siberian species with the latter, Link points out in a brief diagnosis its most significant distinctive features (italics ours): "Imberbis, foliis linearibus longe acutis *apice subfalcatis* paullo longioribus *scapo unifloro, spathis subscariosis* subaequalibus *obtusis* tubo corollae longioribus".

This entirely distinct species is easily differentiated from the closely related *I. ruthenica* Ker-Gawl. with a clearly demarcated distribution range.

11. **I. ventricosa** Pall. Reise, 3 (1776) 712; Maxim. in Bull. Ac. Sci. St.-Pétersb. 26 (1880) 509; Dykes, Gen. Iris (1913) 34; Danguy in Bull. Mus. nat. hist. natur. 20 (1914) 138; B. Fedtsch. in Fl. SSSR, 4 (1935) 516; Kitag. Lin. Fl. Mansh. (1939) 149; Dashnyam in Bot. zh. 50, 11 (1965) 1639. —**Ic.:** Pall. l.c. tab. B, fig. 1; Dykes, l.c.

Described from Dauria (Urulyungui and Argun' interfluvial region near Soktui village). Type in London (BM).

Steppes, rubble and rocky steppe slopes.

IA. **Mongolia:** *East. Mong.* ("Environs de Kailar, steppes, 750 m, No. 1532, June 23, 1896, Chaff."—Danguy, l.c.; around Khailar, No. 565, June 8, 1951—S.H. Li et al.).

General distribution: East. Sib. (Dauria), Far East (south. Ussur.), Nor. Mong. (Fore Hing.), China (Dunbei: nor.-west.).

Section 2. **Pogoniris** Baker

12. **I. bloudowii** Ledeb. Ic. pl. fl. ross. 2 (1830) 5; ej. Fl. alt. 4 (1833) 331; Maxim. in Bull. Ac. Sci. St.-Pétersb. 26 (1880) 533; id. in Acta Horti Petrop.

6, 2 (1880) 498; Dykes, Gen. Iris (1913) 138; Ugrinsky in Beih. Feddes repert. 14 (1922) 18; Kryl. Fl. Zap. Sib. 3 (1929) 670; B. Fedtsch. in Fl. SSSR, 4 (1935) 550; Fl. Kazakhst. 2 (1958) 245. —*I. flavissima* L. α. *umbrosa* Bge. in Ledeb. Fl. alt. 1 (1929) 59. —**Ic.:** Ledeb. Ic. pl. fl. ross. 2, tab. 101; Gartenfl. 29, tab. 1020.

Described from Altay. Type in Leningrad.

Forest and subalpine meadows, meadow slopes of mountains, along coastal meadows.

IIA. **Junggar:** *Cis-Alt.* (Kran river, June 23, 1903—Gr.-Grzh.); *Tarb.* (Hochebene Saur-Altgalten, May 20-June 28, 1876—Pewzow; Saur mountain range, south. slope, Karagaitu river valley, Bain-Tsagan creek valley on right bank, forb subalpine meadow near upper boundary of forest, June 23; same locality, subalpine meadow above larch forest, June 23—1957, Yun.; nor.-west of Khobuk-Saira, Chaganobo hill, on exposed slope, June 22, 1959—Lee et al.); *Jung. Alatau* (Khorgos, 1500-1800 m, May 15, 1879—A. Reg.; Toli district, Albakzin mountains, on slope, No. 2549, Aug. 6, 1957—Kuan); *Tien Shan* (nor.: Talki, July [16]; Sairam lake, south. bank, July 20 and 23—1877, A. Reg.; plateau at sources of Dzhirgalan and Pilyuchi, 1800 m, April 24, 1879—A. Reg.; east.: in forest belt on nor. slope [under Koshety-Daban pass], June 11, 1877—Pot.; nor. slope in larch forest, humus soil, June 2, 1879—Przew.).

General distribution: Jung.-Tarb., Nor. Tien Shan; West. Sib. (Altay).

Note. References of Ugrinsky, Fedtschenko and Kitagawa (Kitagawa, l.c.) to the distribution of this iris in Mongolia, East. Siberia, Far East and Manchuria are erroneous and should be applied to *I. flavissima* Pall. *I. bloudowii* Ledeb. is easily differentiated from *I. flavissima* Pall. in inflated obtuse spathe, veins violet along margins, distant and clearly distinct, joined with transverse anastomoses.

13. **I. flavissima** Pall. Reise, 3 (1776) 715; Turcz. Fl. baic.-dahur. 2, 2 (1856) 197; Maxim. in Bull. Ac. Sci. St.-Pétersb. 26 (1880) 530, p.p., quoad pl. sibir.; Dykes, Gen. Iris (1913) 137, p.p., quoad pl. sibir.; Kryl. Fl. Zap. Sib. 3 (1929) 669, excl. syn.; Pavl. in Byull. Mosk. obshch. ispyt. prir., Biol. 38 (1929) 41; B. Fedtsch. in Fl. SSSR, 4 (1935) 545, p.p., quoad pl. sibir. and excl. syn.; Kitag. Lin. Fl. Mansh. (1939) 147; Grub. Konsp. fl. MNR (1955) 96; Fl. Kazakhst. 2 (1958) 244; Grub. in Novosti sist. vyssh. rast. 6 (1970) 33. —*I. flavissima* ssp. *transuralensis* Ugrinsky in Tr. Obshch. ispyt. prir. Khar'k. univ. 44 (1911) 305; id. in Beih. Feddes repert. 14 (1922) 16. —*I. bloudowii* auct. non Ledeb.: Kitag. l.c. 147. —*I. humilis* Georgi sensu Bobrov in Bot. mat. (Leningrad) 20 (1960) 6. —**Ic.:** Gmel. Fl. sibir. 1 (1747) tab. V, fig. 2; Ugrinsky, l.c. (1922) tab. 1, fig. sin., tab. III, fig. B, tab. IV, fig. 2.

Described from Transbaikal. Type in London (BM) ?

Rubble, rocky and turfed steppe slopes and talus, coastal meadows and pebble beds, among scrubs and on sand.

IA. **Mongolia:** *East. Mong.* (Khailar town region, Nanlintun' village, in scrubs on sandy flat, No. 608, June 10, 1951—S.H. Li).

General distribution: West. Sib. (south. and Altay), East. Sib. (south.), Far East (South.), Nor. Mong., China (nor. and cent. Dunbei).

Note. When studying "type" of *Iris humilis* Georgi, preserved in the herbarium of Moscow University (see M.N. Karavaev, Novosti sist. vyssh. rast. 1973: 335), it was found to be a composite sheet from the personal herbarium of K.S. Trinius labelled *Iris pumila* L.; in fact the sheet contains several complete specimens of this European species. Among them are, however, 2 tiny specimens of *Iris flavissima* Pall. which were neither identified nor marked. The Latin text on the herbarium label—species name and note—was hand-written by K.B. Trinius, as determined by a comparison of handwriting. More importantly, he himself recorded the locality ("ad Baical"). Thus, this herbarium sheet could hardly be credited to I. Georgi, let alone considered the type of *I. humilis* Georgi.

Insofar as the diagnosis of this species is concerned (I.G. Georgi, Bemerkungen..., 1775: 196), even after due consideration of Georgi's explanation (see G.A. Peshkova, Novosti sist. vyssh. rast. 1975: 135), it is impossible to establish with certainty the species he had in view—*I. flavissima* Pall. or *I. potaninii* Maxim. His explanation "Caulis...saepius biflorus, foliis duplo longioribus") is contradictory since leaves are twice longer than stem, which at once is characteristic of *I. potaninii* Maxim.; in *I. flavissima* Pall., they are shorter than stem or equal and concomitantly the peduncle invariably 2-flowered.

14. I. goniocarpa Baker in Gard. Chron. 2 (1876) 710; id. in J. Linn. Soc. (London) Bot. 16 (1877) 145; ej. Handb. Irid. (1892) 24; Hook. f. Fl. brit. India, 6 (1894) 274; Dykes, Gen. Iris (1913) 133.—*I. gracilis* Maxim. in Bull. Ac. Sci. St.-Pétersb. 26 (1880) 527; Forbes and Hemsley, Index Fl. Sin. 3 (1903) 82; Rehder in J. Arn. Arb. 14 (1933) 7.

Described from Himalayas (Sikkim). Type in London (K).

Alpine meadows, meadow slopes of mountains and river valleys, riverbanks and gorge floors as well as among scrubs; 2700–4500 m alt.

IIIA. **Qinghai:** *Nanshan* (South Tetung mountain range, on grassy slopes of mountains, usually on humus clay soil, May 10, 1873—Przew.; An'si, around Kha-Daban, May 3; South Tetung mountain range, Shin'-Chen', May 11—1890, Gr.-Grzh.); *Amdo* (Syan'sibei mountain range, 3350–3660 m, along gorge floor, common, on silty soil, May 21; in Mudzhik mountains, 2900 m, June 17—1880, Przew.; Radja and Yellow river gorges, grassy northern slopes of river valley south of Radja, 3200 m, No. 13983, May 25, 1926—Rock.).

IIIB. **Tibet:** *Weitzan* (pass from Talachu river to Bochu river, 4480 m, July 7, 1884—Przew.; Yantszytszyan river-basin, Donra area on Khichu river [Nyamtso], 3960 m, nor. slope of gorge, on humus in willow thickets, July 16, 1900—Lad.).

General distribution: China (North-west, South-west: Kam), Himalayas (east.).

Note. The distinctive features pointed out by Maximowicz differentiating *I. gracilis* Maxim. from *I. goniocarpa* Baker do not in fact exist. They arose out of the inadequate and incomplete description of *I. goniocarpa* cited by Baker since Maximowicz did not study specimens of this species but rather compared his species with Baker's description. Dykes (l.c.) was totally justified in relegating this Maximowicz species to synonyms of *I. goniocarpa* Baker.

15. I. pandurata Maxim. in Bull. Ac. Sci. St.-Pétersb. 26 (1880) 529; Forbes and Hemsley, Index Fl. Sin. 3 (1903) 83. —? *I. ruthenica* auct. non Ker-Gawl.: Diels in Filchner, Wissensch. Ergebn. (1908) 249. —*I. tigridia* auct. non Bunge: Dykes, Gen. Iris (1913) 153, quoad pl. e prov. Kansu.

Described from Qinghai (Nanshan). Type in Leningrad. Plate VI, fig. 5; map 5.

Rocks, steep rocky slopes and precipices, forest (?) belt.

IIIA. **Qinghai:** *Nanshan* (South Tetung mountain range, lower belt, on steep rocks, very uncommon, May 29, 1873—Przew., typus!; valley of Dzhanba [Itel'-Gol] river, April 15; Nimbi-Muren [Xining-Gol] river valley, right bank, on silty-rocky soil, June 18, Laovasya [Lovachen] town, Nimbi-Muren river valley at beginning of gorge, along south. flank, on rocky soil, April 19—1885, Pot.); *Amdo* (along tributary of Baga-Gorgi river, 2900 m, silty gorge cliffs, common, May 8; along Huang He river near Churmyn river estuary, 2600 m, May 18—1880, Przew.; valley of Karym river before Kazhir village, May 5; valley of Lonchzhu-Lunva river, May 8; along Nuryn-Dzhamba [Mudzhik] river, May 11—1885, Pot.).

General distribution: endemic.

Note. Distinct species with stable characteristics and genetically related to *I. potaninii* Maxim. Dykes (l.c.) erroneously treated it as a synonym of *I. tigridia* Bge. for want of adequate herbarium material.

16. **I. potaninii** Maxim. in Bull. Ac. Sci. St.-Pétersb. 26 (1880) 528; Ugrinsky in Beih. Feddes repert. 14 (1922) 19, excl. pl. tibet. et chin.; Kryl. Fl. Zap. Sib. 3 (1929) 668; B. Fedtsch. in Fl. SSSR, 4 (1935) 549; Grub. Konsp. fl. MNR (1955) 97; Popov, Fl. Sr. Sib. 1 (1957) 206. —*I. flavissima β. rupestris* Bge. in Ledeb. Fl. alt. 1 (1829) 60. —*I. pumila* auct. non L.: Pall. Reise, 3 (1776) 715. —*I. tigridia* auct. non Bunge: Hanelt and Davazamè in Feddes repert. 70, 1–3 (1965) 20. —**Ic.:** Ugrinsky, l.c. tab. IV, fig. 3; Popov, l.c. fig. 23, 2.

Described from East. Siberia (Angara river). Type in Leningrad. Plate VI, fig. 3.

Steppes, more often rubble and rocky, slopes of knolls and mountains, rarely in sandy steppes as well as on rocks and talus; up to montane steppe belt and 2700 m.s.m.

IA. **Mongolia:** *Khobd.* (between Ukha and Khatu, July 10, 1870—Kalning; top of Ulan-Daban pass, June 22, 1879—Pot.); *Mong. Alt.* (Dain-Gol lake, June 27, 1903—Gr.-Grzh.); *Cent. Khalkha* (Khukhu-Khoshu, rocky slope, July 24, 1926—Lis.; Kholt area, May 23, 1926—Gus.; 5 km east of Choiren-Ula, rubble feather-grass steppe, Aug. 22, 1940; Choiren-Ula and Ulan-Bator—Sain-Shanda road, forb-grassy dry steppe on granite eluvium, May 28, 1941; Sorgol-Khairkhan hill 150 km from Ulan-Bator along old road to Dalan-Dzadagad, in rock crevices, May 16, 1941; Mandal somon, Sakhil'te-Ula, rocky slopes, May 10; Moltsok-Elisu sand 3–4 km nor. of Moltsok-Khida, hummocky semifixed sand, May 16, 1944—Yun.); *East. Mong.* (Manchuria railway station, steppe, June 5, 1902—Litw.; same locality, steppe slopes of hummocks, May 6, 1908—Kom.; Erdaotszintszy village south of Manchuria station, 800 m, hill slopes, No. 954, July 26, 1951—S.H. Li et al.; Choibalsan somon, Khabirga station region nor. of Enger-Shanda, east. gentle slope of knoll, forb steppe, Aug. 1, 1954; same locality, 11 km nor. of Enger-Shanda, top of knoll, May 28, 1958—Dashnyam); *Val. Lakes* (Telin-Gol and Tatsain-Gol interfluvial region along road, wormwood-forb dry steppe on rubble soil, June 27, 1941—Tsatsenkin); *Gobi-Alt.* (Dundu-Saikhan mountains, west. slope, upper belt, July 5, 1909—Czet.; Gurban-Saikhan, dry slopes in canyons and gullies, 2000–2300 m, No. 123, 1925—Chaney; nor. foothill of Ikhe-Bogdo mountain range, gorge in offshoots, May 16, 1926—Kozlova; Bain-Tsagan mountain range, nor. of Bain-Tukhum area, mountain slopes in upper belt, May 28, 1938—Luk'yanov; nor. slope of Ikhe-Bogdo mountain range, Shishkhid creek valley, 2650 m, south. steep talus slope, June 30; same locality, upper Shishkhid creek valley, along gullies in mountain steppe belt, June 30—1945, Yun.); *East Gobi* (west. extremity of Delger-Khangai mountain range, in pass, feather-grass dry montane steppe, May 10, 1941—Yun.).

General distribution: West. Sib. (Altay: Chui steppe), East. Sib. (Kizyl vicinity, East. Tannu-Ol, Khamar-Daban, Fore Baikal and Transbaikal; far north. find: Barguzin valley near Bodon village, 53°40′ N. lat. and Kurkut bay near Ol 'khon island; far east: Manchuria station), Nor. Mong. (Fore Hubs., Hang., Mong.-Daur.).

Note. Flowers pale, brownish inside.

17. **I. scariosa** Willd. ex Link in Spreng., Schrad. and Link, Jahrb. 1, 3 (1820) 71; Maxim. in Bull. Ac. Sci. St.-Pétersb. 26 (1880) 534; Dykes, Gen. Iris (1913) 178; Kryl. Fl. Zap. Sib. 3 (1929) 671; B. Fedtsch. in Fl. SSSR, 4 (1935) 550; Fl. Kazakhst. 2 (1958) 245. —*I. glaucescens* Bge. in Ledeb. Fl. alt. 1 (1829) 58; ej. Ic. pl. fl. ross. 2 (1830) tab. 102. —*I. eulefeldii* Rgl. in Acta Horti Petrop. 5, 2 (1878) 633 and in Gartenfl. 27 (1878) 325. —*I. glaucescens* var. *eulefeldii* Maxim. in Acta Horti Petrop. 6, 2 (1880) 498. —Ic.: Ledeb. Ic. pl. fl. ross. 2, tab. 102; Gartenfl. 27, tab. 954; Bot. Mag. 112, tab. 6902; Fl. Kazakhst. 2, tab. XXII, fig. 4.

Described from Caspian coast. Type in Berlin (B).

Steppe, rocky and rubble slopes of knolls, ridges, mountains and river valleys, solonetz sand; in mountains up to 2700 m alt.

IIA. Junggar: *Cis-Alt.* (Bugotar area between Burchum and Kran rivers, June 11, 1903—Gr.-Grzh.; 15–20 km nor.-west of Shara-Sume along Kran river, scrub meadow steppe, July 7, 1959—Yun.). *Tien Shan* (Kul'dzha and east.; far east. locality: in steppe valley [Sochzhan river] nor. of mountain pass [Koshety-Daban], Tien Shan mountain range, sandy soil, June 12, 1877—Pot.); *Zaisan* (Ch. Irtysh river, left bank, Dzhelkaidar, hummocky sand, June 9, 1914—Schischk.).

General distribution: Aralo-Casp., Fore Balkh., Jung.-Tarb.; West. Sib. (south. and Altay).

Note. 1. The Herbarium of the Komarov Botanical Institute of the Academy of Sciences, USSR, has a single specimen of this species from the Caspian coast—"Astrachan, Blum"—cited by Maximowicz (l.c.), but the latter mentions 2 more Karelin's specimens from the east coast of the Caspian and Volga estuary. These, however, could not be found. There is yet another specimen in F. Fischer's herbarium with the inscription "Turcomania". More recent collections of *I. scariosa* from the Caspian region have not been found.

2. *I. eulefeldii* Rgl., no different from common *I. scariosa* Willd. but for large size of all plant parts, could be treated as a large montane variety of this species, as was done by Maximowicz—*I. scariosa* Willd. var. *eulefeldii* (Rgl.) Maxim. l.c. This variety is found along with the typical in Kul'dzha region of Tien Shan and in Junggar Alatau. Its stem is up to 30 cm tall and its capsule 6 cm long, leaves twice broader and longer than in typical variety.

18. **I. thoroldii** Baker in Hooker, Ic. pl. 24 (May 1894) tab. 2302; id. in Hemsley in J. Linn. Soc. (London) Bot. 30 (1894) 118, 129, 139; Hemsley, Fl. Tibet. (1902) 199. —*I. potaninii* auct. non Maxim.: Forbes and Hemsley, Index Fl. Sin. 3 (1903) 83; Diels in Filchner, Wissensch. Ergebn. (1908) 249; Dykes, Gen. Iris (1913) 154; Rehder in J. Arn. Arb. 14 (1933) 7. —*I. tigridia* auct. non Bunge: Danguy in Bull. Mus. nat. hist. natur. 17 (1911) 449. —**Ic.** Hooker, l.c.; Dykes, l.c.

Described from Tibet (Chang Tang). Type in London (K). Plate VI, fig. 4; map 5.

Turfed and exposed clayey-rocky slopes, among rocks, clayey areas and sandy and pebble banks and shoals of rivers, 2750–5500 m alt.; forms compact, often circular mats.

IC. Qaidam: *montane.* (Sarlyk-Ula mountains, Elisten-Kuku-Bulak area, 3350–3650 m, rocky soil, May 11; same locality, 3350 m, nor. slope, May 20—1895, Rob.).

IIIA. Qinghai: *Nanshan* (Shuiguan'chzhatszy post in Lagi river valley, 2890 m, May 28; nor. slope of Lagin-Daban pass, May 30—1886, Pot.; South Kukunor mountain range, south. slope, 3350 m, on loess, May 6, 1895—Rob.; "Yong-Ngan [Xining district], 3000 m, July 7, 1908, Vaillant"—Danguy, l.c.); *Amdo* (Baga-Gorgi river gorge, 2900 m, on grassy slope, common, May 3; along Baga-Gorgi river tributary, among rocks on silty soil, dispersed, May 9—1880, Przew.; "Siansibei, Labtse, July 7, 1904, Filchner"—Diels, l.c.; "Radja and Yellow river gorges; Jupar Range, 1924–1927, J. Rock"—Rehder, l.c.).

IIIB. Tibet: *Chang Tang* ("Top of pass, at 17,800 ft [5430 m], 116 bis, 1891—Thorold., typus!; Sharakuyi-gol, N. lat. 35°50′, E. long 93°27′, hill slope at 13,800 ft [4200 m], May 29, 1892—Rockhill; N. lat. 35°16′, E. long. 91°, 16,000 ft [4970 m], Aug. 8, 1896—Wellby and Malcolm"—Baker l.c.); *Weitzan* (Nomokhun river gorge, along brook among stones, scattered, May 23; in Nomokhun river valley, 4250 m, on moist sand, May 24; south. slope of Huang He and Yangtze watershed, June 11; in mountains along Dychu river course, June 20; left bank of Yangtze river [Dychu], 3950 m, June 29—1884, Przew.; nor. slope of Burkhan-Budda mountain range, Nomokhun river gorge, 3650 m, May 18; same locality, in alpine belt, 4250 m, on humus, May 18; south. slope of Burkhan-Budda mountain range, 4250 m, bed and mountain descents, May 30; Alak-Nor lake, 3650–3950 m, bald loose clayey areas, May 30; Russkoe lake, Ekspeditsii lake, stream and Dzhagyn-Gol, 3950–4250 m, June 15—1900, Lad.; Yantszytszyan basin, Tszergen area, 4100 m, sunny side of mountains and along gorge floor, clayey-rocky soil and humus, May 14; same locality, along Gorin'chu river valley, 4250 m, clay and humus, May 17; Huang He basin, upper Sergchu river, 4250 m, clayey soil, May 23—1901, Lad.).

General distribution: China (North-West: Gansu, Gonpa-Lonva brook above Labran; South-west: Kam, Yangtze basin, Ombunda area along Dzachu river).

Note. This endemic Tibetan species is readily distinguished from the closely related species *I. tigridia* Bge. and *I. potaninii* Maxim. in obtuse or shortly acuminate light green leaves and abundant crispate coarse-fibrous sheath remnants of dead leaves in mat. In fresh plants, colour of flowers varies from pale and yellow, through light and dull lilac to lilac and dark violet (as shown on labels). Sometimes, flowers are coloured differently in the same mat. Flowers fragrant. Dykes in his monograph (l.c.), for some incomprehensible reason, confused this species with *I. potaninii* Maxim. and cited it among synonyms of the latter, but his list of reports does not contain a single genuine

I. potaninii Maxim. from its distribution range, including even authentic specimens from East. Siberia cited by Maximowicz. Moreover, he noted on one of the authentic specimens: "*I. tigridia* Bge! Die echte *I. potaninii* unterscheidet sich durch die viel stumpferen Blätter. Det. W.R. Dykes". So great was his conviction that Tibetan specimens are represented as genuine *I. potaninii* Maxim. in spite of the fact that this species was known as East. Siberian! The illustration reproduced by him on the same page in the monograph belongs not to *I. potaninii* Maxim. but to *I. thoroldii* Baker and demonstrates very clearly the distinctive characteristics of this species—obtuse leaves and crispate fibrous sheath remnants.

19. **I. tigridia** Bge. in Ledeb. Fl. alt. 1 (1829) 60 and Ic. pl. fl. ross. 4 (1833) 14; Turcz. Fl. baic.-dahur. II, 2 (1856) 198; Maxim. in Bull. Ac. Sci. St.-Pétersb. 26 (1880) 529; Dykes, Gen. Iris (1913) 153, excl. *I. pandurato* Maxim. p.p.; Danguy in Bull. Mus. nat. hist. natur. 20 (1914) 137; Kryl. Fl. Zap. Sib. 3 (1929) 668; Pavl. in Byull. Mosk. obshch. ispyt. prir. Biol. 38 (1929) 41; B. Fedtsch. in Fl. SSSR, 4 (1935) 549; ?Kitag. Lin. Fl. Mansh. (1939) 149; Grub. Konsp. fl. MNR (1955) 97; Fl. Kazakhst. 2 (1958) 244. —**Ic.:** Ledeb. Ic. pl. fl. ross. 4, tab. 342; Fl. SSSR, 4, Plate XXIV, fig. 6.

Described from Altay (Charysh river). Type in Leningrad.

Rubble and rocky steppe slopes of knolls and mountains, sandy steppes, talus and rocks.

IA. Mongolia: *Cent. Khalkha* (Bain-Undur somon, Dagan-Dele area along road to Arbai-Khere from Ulan-Bator, upper part of rocky slope of mud cone, June 19, 1952—Davazamč); *East. Mong.* (3 versts nor.-west of Tsagan-Balgasu, on mountain, in rocky soil, May 1, 1831—Ladyzh.; Manchuria railway station, steppe, June 5, 1902—Litw.; 10 km nor. of Yangao town, Yunmen'-Shan' mountain near Yan'tszya village, rocky slopes, 1500 m, May 19, 1957—Petr.); *?Khesi* (67 km nor. of Lan'chzhou town, wormwood-forb-saltwort semidesert, June 29, 1957—Petr.).

General distribution: West. Sib. (Altay), East. Sib. (Transbaikal and Dauria), Nor. Mong., China (Dunbei: east. slope of B. Hinggan in Tsitsikara region).

Section 3. **Pardanthopsis** (Hance) Baker

20. **I. dichotoma** Pall. Reise, 3 (1776) 712; Maxim. in Bull. Ac. Sci. St.-Pétersb. 26 (1880) 540; Franch. Pl. David. 1 (1884) 298; Forbes and Hemsley, Index Fl. Sin. 3 (1903) 81; Palibin in Tr. Troitskosavsko-Kyakht. otdelen. RGO, 7, 3 (1904) 41; Dykes, Gen. Iris (1913) 96; B. Fedtsch. in Fl. SSSR, 4 (1935) 530; Kitag. Lin. Fl. Mansh. (1939) 147; Grub. Konsp. fl. MNR (1955) 96. —*Pardanthus dichotomus* Ledeb. Fl. Ross. 4 (1852) 106; Turcz. Fl. baic.-dahur. II, 2 (1856) 199. —**Ic.:** Pall. l.c. tab. A, fig. 2; Bot. Reg. 3, tab. 246; Bot. Mag. tab. 6428; Fl. SSSR, 4, Plate XXXII, fig. 1.

Described from Dauria (between Ingoda and Argun'). Type in London (BM).

Feather-grass-forb and forb-tansy steppes on slopes of knolls, sandy steppes and among steppe scrubs.

IA. Mongolia: *Cent. Khalkha* (Durkhe [Darkha]-Arshan, Aug. 13, 1927—Terekhovko); *East. Mong.* (far south-west. report or locality Choibalsan somon, 12 km west of Khabirga station, Aug. 12, 1954—Dashnyam; 3 versts—nor.-east of Tsagan-Balgasu, June 25, 1831—Ladyzh.; Kalgan gorge, Aug. 23, 1898—Zab.; south. slope, Muni-Ula mountain range, on bald clayey slope, July 19, 1871—Przew.); *Alash. Gobi* (Alashan mountain range, Baisy monastery, 2100 m, juniper thickets, July 6, 1957—Petr.); *Ordos* (east.: nor. of Narin-Gol river, on shifting dunes, Sept. 10, 1884—Pot.; Huang He river valley 30 km south of Dalat town, Chzhandanchzhao village vicinity, Aug. 10, 1957—Petr.).

IIIA. Qinghai: *Nanshan* (Sanchuan', Itel'-Gol area, April 1885—Pot.); *Amdo* (Dzhakhar mountains on Mudzhik river, July 5, 1890—Gr.-Grzh.).

General distribution: East. Sib. (Daur.), Nor. Mong. (Mong.-Daur.—far west. report: Bain-Dung somon, Dzun-Naran area; Fore Hing.), China (Dunbei, North?, North-west, South-west: Hubei).

Section 4. Articulata Dykes

21. **I. kolpakowskiana** Rgl. in Acta Horti Petrop. 5 (1877) 263; Maxim. in Bull. Ac. Sci. St.-Pétersb. 26 (1880) 504; Dykes, Gen. Iris (1913) 228; B. Fedtsch. in Fl. SSSR, 4 (1935) 556; Fl. Kirgiz. 3 (1951) 126. —*Xiphion kolpakowskianum* Baker in Bot. Mag. (1880), tab. 6489; Fl. Kazakhst. 2 (1958) 247. —**Ic.:** Gartenfl. 27, tab. 939; Fl. SSSR, 4, Plate XXXIV, fig. 5; Fl. Kazakhst. 2, Plate XXI, fig. 5.

Described from Tien Shan (Alma Ata district). Type in Leningrad.

Forb and rocky steppe slopes, from foothill plains to 1000 m.s.m.

IIA. Junggar: *Tien Shan* (Pilyuchi; Toguztorau west of Kul'dzha, April 22–24, 1879—A. Reg.).

General distribution: Nor. Tien Shan; Mid. Asia (West. Tien Shan).

Family 28. ORCHIDACEAE Juss.

1. Achlorophyllous saprophyte, leafless. Flowers greenish, small; labellum without spur .... 2.
+ Plant green, with normally developed caulous or only radical leaves .... 3.
2. Rhizome branched, coral-like, without fibrous roots. Labellum not longer than other perianth lobes, oblong, undivided ....
....................................3. **Corallorhiza** Chatel. (*C. trifida* Chatel.).
+ Rhizome shortened and densely covered with upcurved funiform roots (like bird's nest). Labellum exserted, twice longer than other perianth lobes, deeply bilobate ....
.... 5. **Neottia** Guett. [*N. camtschatea* (L.) Reichb. f.].
3. Flowers very large, 3–8 cm in diam., bright coloured and with large saccate inflated labellum, terminal, solitary or rarely 2–3.

Large rhizomatous plant with 2–5 elliptical sessile caulous leaves .... 1. **Cypripedium** L.

+ Flowers generally small and medium in size, not more than 2 cm in diam., invariably in racemose inflorescence and usually numerous; labellum saccate or inflated .... 4.

4. Inflorescence axis spirally twisted; flowers small, purple, in dense spicate raceme on tall stem. Leaves linear-lanceolate or linear, radical, 2–4. Roots thickened, funiform .... 7. **Spiranthes** Rich.

+ Inflorescence axis not twisted .... 5.

5. Stem with normally developed leaves .... 6.

+ Stem leafless or only with small bract-like or scale-like leaves; radical leaves developed .... 11.

6. Leaves 2, opposite, in midportion of stem, sessile, broadly cordate. Rhizome slender, creeping. Flowers very small, greenish; labellum bilobed, without spur; bracts very small, scale-like ....
.... 4. **Listera** R. Br.

+ Leaves alternate, oblong-ovate to linear-lanceolate and linear. Bracts large, graminaceous .... 7.

7. Plant with undivided ovate or palmately lobed bulbs. Labellum with spur, not narrowed, flat. Capsule erect .... 8.

+ Plant with creeping rhizome. Labellum without spur, transversely narrowing into 2 parts. Capsule nutant or pendent ....
.... 6. **Epipactis** Zinn.

8. Flowers yellowish-green, with very dark narrowly obcuneate labellum and short saccate spur. Leaves 2–4 (5), oblong-ovate to lanceolate-elliptical. Bulbs bipartite, oblong ....
.... 12. **Coeloglossum** C. Hartm. [*C. viride* (L.) C. Hartm.].

+ Flowers lilac-pink or pink, with broad labellum and long cylindrical or filiform spur. Leaves numerous or 2 (rarely 3), elliptical or lanceolate, sublinear .... 9.

9. Inflorescence regular, many-flowered, generally compact. Leaves numerous, elliptical, lanceolate or linear. Bulbs 2–6-lobed ...... 10.

+ Inflorescence secund, lax, with 2–8 lilac-purple flowers. Small plant, 10–20 cm tall, with slender stem, bearing 2 (rarely 1) oblong leaves near middle, and undivided pea-size bulb ....
.... 17. **Ponerorchis** Reichb. f. [*P. pauciflora* (Lindl.) Ohwi].

10. Flowers lilac-pink, monochromatic, with slender spur falcately decurved; spur 3–4 times longer than flower and 2 times longer than ovary; viscidia naked. Leaves linear-lanceolate to linear. Bulbs 4–6-lobed ....
.... 15. **Gymnadenia** R. Br. [*G. conopsea* (L.) R. Br.].

+ Flowers with dark-coloured pattern on labellum and with cylindrical or conocylindrical, straight or faintly curved spur, not longer than ovary; viscidia in pouch. Leaves elliptical to linear-lanceolate. Bulbs 2–4–6-lobed .... 18. **Orchis** L.

11. Leaves 4–8, ovate, on short petiole, with distinct reticulate venation formed of 5 oblong veins joined with numerous transverse anastomoses. Flowers small, white, in secund dense raceme. Small plant with long creeping rhizome ....
.... 8. **Goodyera** R. Br. [*G. repens* (L.) R. Br.].

+ Leaves 1–4, obovate to linear, without distinct venation .... 12.

12. Plant with creeping slender rhizome and solitary elliptical or lanceolate radical leaf. Flowers few, distant, in sparse inflorescence .... 16. **Galeorchis** Rydb.

+ Plant with bulb or thickened fusiform or funiform roots .... 13.

13. Plant with solitary globose or ovoid bulb. Leaves elliptical to linear, acute or acuminate .... 14.

+ Plant with thick funiform or fusiform roots. Leaves obovate or oblong-obovate, obtuse. Flowers greenish-white; labellum undivided with cylindrical spur .... 17.

14. Flowers violet-pink, in rather lax secund many-flowered raceme; all perianth lobes, apart from labellum, agglutinated into rather narrow galea. Leaves in pairs nearly opposite, elliptical ....
.... 11. **Neottianthe** Schlecht. [*N. cucullata* (L.) Schlecht.].

+ Flowers yellowish-green, very small, in regular raceme; perianth lobes free .... 15.

15. Bulb covered with white scale-like sheaths. Leaves in pairs, elliptical. Raceme slender, with distant flowers ....
.... 2. **Malaxis** Soland. [*M. muscifera* (Lindl.). Grub.].

+ Bulbs without scales, dark coloured. Leaves 1–4, elliptical to linear .... 16.

16. All perianth lobes folded together into a cup; labellum 3-lobed or undivided, without or with short saccate spur. Leaves elliptical to linear .... 10. **Herminium** Guett.

+ Perianth lobes unfolded; labellum 3-lobed, with cylindrical spur. Leaves linear ....
.... 9. **Habenaria** Willd. (*H. spiranthiformis* Ames et Schlechter).

17. Small plant, 7–20 cm tall, with 1 oblong-obovate leaf and thickened funiform roots. Flowers small, in few-(3–6)-flowered raceme .... 14. **Lysiella** Rydb. [*L. oligantha* (Turcz.) Nevski].

+ Plant 20–40 cm tall with 2 nearly opposite obovate leaves and fusiform roots. Flowers large, with slender spur falcately upcurved, in many-flowered raceme ....
.... 13. **Platanthera** Rich. (*P. freynii* Kranzl.).

1. **Cypripedium** L.

Sp. pl. (1753) 951.

1. Plant with 2 leaves proximate midstem, blackish on drying. Rhizome slender, funiform. Flower solitary; labellum shorter than obtuse perianth lobes, raspberry-red with white spots ....
.... **C. guttatum** Sw.

+ Plant with 3–4 (5) alternate leaves, not blackish on drying. Rhizome thick. Flowers solitary, sometimes 2–3, with acute perianth lobes and monochromatic labellum .... 2.

2. Labellum light yellow, shorter than creamish-brown lanceolate perianth lobes, of which lateral narrower than rest and slightly contorted axially. Flowers solitary or 2–3 .... **C. calceolus** L.

+ Labellum longer than lateral ovate-lanceolate flat perianth lobes or almost equal and same in colour. Flowers solitary, rarely in pairs .... 3.

3. Perianth violet-pink; staminode white, violet-speckled ....
.... **C. macranthon** Sw.

+ Perianth purple-red; staminode yellow ....
.... 1. **C. fasciolatum** Franch.

1. **C. fasciolatum** Franch. in J. Bot. (Paris) 8 (1894) 232; Forbes and Hemsley, Index Fl. Sin. 3 (1903) 64; Schlechter in Feddes repert., Beih. 4 (1919) 81; id. in Acta Horti Gothoburg. 1 (1924) 127; Walker in Contribs. U.S. Nat. Herb. 28 (1941) 604. —?*C. tibeticum* auct. non King ex Rolfe (1892): Schweinf. in J. Arn. Arb. 10 (1929) 170.

Described from China (Sichuan). Type in Paris (P).

Coniferous and mixed forests, coastal, forest and alpine meadows.

IIIA. Qinghai: *Nanshan* (South Tetung mountain range, July 8, 1872—Przew.); *Amdo* ("Radja and Yellow river gorges; grassy northern slopes of river valley south of Radja, alt. 3200 m, No. 13,989, May 1926 (flowers red to purple); alpine meadows of mountains opposite Radja, alt. 3500 m, No. 14165, June 1926—J. Rock"—Schweinf. l.c.).

General distribution: China (North-west: south. Gansu; South-west).

Note. Judging from the colour of flowers (from red to purple) recorded by Schweinfurth, l.c.), J. Rock's specimens pertain to this species and not to *C. tibeticum* King ex Rolfe, since flowers of the latter are variegated—labellum purplish-brown and perianth lobes albescent.

C. calceolus L. Sp. pl. (1753) 951; Forbes and Hemsley, Index Fl. Sin. 3 (1903) 64; Danguy in Bull. Mus nat. hist. natur. 20 (1914) 136; Schlechter in Feddes repert., Beih. 4 (1919) 80; Kryl. Fl. Zap. Sib. 3 (1929) 679; Nevski in Fl. SSSR, 4 (1935) 598; Kitag. Lin. Fl. Mansh. (1939) 150; Grub. Konsp. fl. MNR (1955) 97; Fl. Kazakhst. 2 (1958) 255. —**Ic.:** Reichb. Ic. Fl. Germ. 13–14, tab. 144.

Described from Eurasia. Type in London (Linn.).

Deciduous and mixed sparse forests, forest fringes and meadows, scrubs.

Found in regions adjoining Nor. Mongolia and China.

General distribution: Europe, Balk., West. and East. Sib., Far East, Nor. Mong. (Hent., Mong.-Daur.), China (Dunbei, North), Korean peninsula.

C. guttatum Sw. in Kongl. Sv. Vet. Ac. Handl. 21 (1800) 251; Franch. Pl. David. 1 (1884) 295; Forbes and Hemsley, Index Fl. Sin. 3 (1903) 64; Danguy in Bull. Mus. nat. hist. natur. 20 (1914) 136; Schlechter in Feddes repert., Beih. 4 (1919) 81; ibid. Beih. 12 (1922) 326; id. in Acta Horti Gothoburg. 1 (1924) 128; Kryl. Fl. Zap. Sib. 3 (1929) 675; Nevski in Fl. SSSR, 4 (1935) 596; Kitag. Lin. Fl. Mansh. (1939) 151; Grub. Konsp. fl. MNR (1955) 97; Fl. Kazakhst. 2 (1958) 255. —Ic.: Reichb. Ic. Fl. Germ. 13–14, tab. 143.

Described from Siberia. Type in Stockholm (S).

Larch and birch-larch forests, birch groves, forest fringes.

Found in regions adjoining Nor. Mongolia and China (Fore Hing., Arshan town vicinity).

General distribution: Europe (nor. half of European USSR), West. and East. Sib., Far East, Nor. Mong., China (Dunbei, North, South-west?), Korean peninsula.

C. macranthon Sw. in Kongl. Sv. Vet. Ac. Handl. 21 (1800) 251; Franch. Pl. David. 1 (1884) 295; Forbes and Hemsley, Index Fl. Sin. 3 (1903) 66; Danguy in Bull. Mus. nat. hist. natur. 20 (1914) 137; Schlechter in Feddes repert., Beih. 4 (1919) 83; ibid. Beih. 12 (1922) 327; id. in Acta Horti Gothoburg. 1 (1924) 128; Kryl. Fl. Zap. Sib. 3 (1929) 677; Nevski in Fl. SSSR, 4 (1935) 598; Kitag. Lin. Fl. Mansh. (1939) 151; Grub. Konsp. fl. MNR (1955) 97; Fl. Kazakhst. 2 (1958) 256. —Ic.: Reichb. Ic. Fl. Germ. 13–14, tab. 146.

Described from Siberia. Type in Stockholm (S).

Sparse larch, birch-larch and birch forests, forest meadows fringes.

Found in regions adjoining Nor. Mongolia and China (Fore Hing., Arshan town vicinity).

General distribution: Europe (nor. European USSR), West. and East. Sib., Far East, Nor. Mong., China (Dunbei, North), Korean peninsula, Japan (nor.).

2 **Malaxis** Soland. ex Sw.

Prodr. Veg. Ind. Occid. (1788) 119. —*Microstylis* Nutt. Gen. Amer. 2 (1818) 196. —*Dienia* Lindl. in Bot. Regist. (1824) tab. 825.

1. **M. muscifera** (Lindl.) Grub. comb. nova hoc. loco. —*Dienia muscifera* Lindl. Gen. a. sp. Orchid. (1830) 23. —*Microstylis muscifera* Ridl. in J. Linn. Soc. (London) Bot. 24 (1888) 333; Hook, f. Fl. Brit. Ind. 5 (1890) 689.

Described from Himalayas (Nepal). Type in London (K).

Shady moss-covered forests, moist shaded rocks.

IIIA. **Qinghai:** *Nanshan* (along Yusun-Khatyma river, 2700 m, on rocks, rare, July 24; South Tetung mountain range, in forest, Aug. 4—1880, Przew.).
General distribution: Himalayas.

Note. Species very closely related to common *M. monophyllos* (L.) Sw. found in Nor. Mongolia and China but invariably 2-leaved and thicker and coarser.

3. Corallorhiza Chatel.
Specim. inaug. Corall. (1760) 1.

1. **C. trifida** Chatel. Specim. inaug. Corall. (1760) 8; Kraenzl. in Feddes repert., Beih. 65 (1931) 91; Nevski in Fl. SSSR, 4 (1935) 608; Fl. Kirgiz. 3 (1951) 135; Grub. Konsp. fl. MNR (1955) 98; Fl. Kazakhst. 2 (1958) 257. — *C. neottia* Scop. Fl. carniol. 2 (1772) 207; Kryl. Fl. Zap. Sib. 3 (1929) 717. — *C. innata* R. Br. in Ait. Hort. Kew., ed. 2, 5 (1813) 209; Forbes and Hemsley, Index Fl. Sin. 3 (1903) 9; Schlechter in Feddes repert., Beih. 4 (1919) 222; id. in Acta Horti Gothoburg. 1 (1924) 151. —*Ophrys corallorhiza* L. Sp. pl. (1753) 345. —**Ic.:** Reichb. Ic. Fl. Germ. 13, tab. 138; Fl. SSSR, 4, Plate XXXVII, fig. 11; Fl. Kazakhst. 2, Plate XXIV, fig. 3.
Described from Nor. Europe. Type in London (Linn.).
Humid coniferous forests.

IIA. **Junggar:** *Tien Shan* (Syata-Ven'tsyuan', in forest along river-bank, 2200–2900 m abs. height, No. 1419, Aug. 13, 1957—Shen-Tyan').
General distribution: Jung.-Tarb. (Jung. Alatau), Nor. Tien Shan (Transili Alatau); Europe, Balk. (west.), Caucasus, West. Sib., East. Sib., Far East, Nor. Mong. (Fore Hubs., Hent.), China (North?, South-west), Nor. America (nor.).

4. Listera R. Br.
in Ait. Hort. Kew., ed. 2, 5 (1813) 201.

1. Raceme many-flowered, long; perianth 2.5–3 mm long; labellum thrice longer than perianth, narrow, with acute notch slightly enlarged and bifurcate at end with short acute lobes. Stem with 1–2 leafy bracts below inflorescence .... 1. **L. puberula** Maxim.

\+ Raceme few-(1–3?)-flowered; perianth 3.5 mm long; labellum slightly longer than perianth, 4.5–5 mm long, abruptly enlarged spatulately from middle, up to 2 mm broad and notched at tip into 2 orbicular lobes. Stem without leafy bracts below inflorescence .... 2. **L. tianschanica** Grub.

1. **L. puberula** Maxim. in Bull. Acad. Sci. St.-Pétersb. 29 (1883) 204; Forbes and Hemsley, Index Fl. Sin. 3 (1903) 41; Schlechter in Feddes repert., Beih. 4 (1919) 142; Nevski in Tr. Bot. inst. An SSSR, 1 ser., 2 (1936) 118.
Described from Qinghai (Nanshan). Type in Leningrad.
Shady and moss-covered coniferous forests.

IIIA. Qinghai: *Nanshan* (North Tetung mountain range, lower part of forest belt, about 2300 m, in coniferous forest, among mosses, rare, Aug. 11, 1880—Przew., typus!).

General distribution: endemic.

2. L. tianschanica Grub. sp. nova.

Planta habitu et dimensione *L. cordata* R. Br. valde similis est sed labello spathulato sepalis vix sesquiplo longiore apice breviter bifido lobulis rotundatis, scapo et racemo glanduloso-puberulo facile distinguitor.

Typus: Songaria, Tianschan: ad ripam fl. Ulan-Usu 28 km ad meridioorientem ab opp. Njutzuanjtza in regione silvestri in declivio oriente-boreali, July 18, 1957 —K.C. Kuan, in Herb. Inst. Bot. Ac. Sci. URSS (Leningrad) conservatur.

Plate V, fig. 3.

IIA. Junggar: *Tien Shan* (along bank of Ulan-Usu river 28 km south-east of Nyutsyuan'tsz, in forest belt, on slope of nor.-east. exposure, July 18, 1957—Kuan, typus!).

5. Neottia Guett.
in Mém. Hist. Ac. Sci. Paris 1750 (1754) 374.

1. **N. camtschatea** (L.) Reichb. f. Ic. Fl. Germ. 13–14 (1851) 146; Schlechter in Feddes repert., Beih. 4 (1919) 140; Kraenzl. in Feddes repert., Beih. 65 (1931) 78; Nevski in Fl. SSSR, 4 (1935) 619; Fl. Kirgiz. 3 (1951) 135; Grub. Konsp. fl. MNR (1955) 98; Fl. Kazakhst. 2 (1958) 260; Fl. Tadzh. 2 (1963) 400. —*N. camtschatica* Spreng. Syst. 3 (1826) 707; Forbes and Hemsley, Index Fl. Sin. 3 (1903) 40; Kryl. Fl. Zap. Sib. 3 (1929) 711. — *Ophrys camtschatea* L. Sp. pl. (1753) 948. --Ic.: Reichb. Ic. Fl. Germ. 13, tab. 126; Fl. SSSR, 4, Plate XXXVIII, fig. 6.

Described from Siberia. Type in London (Linn.).

Shady humid coniferous forests, sometimes in humid scrubs in forest belt of mountains.

IIA. Junggar: *Tien Shan* (Kash river valley, Arystyn, 2400 m, July 27, 1879—A. Reg.; Manas river-basin, left bank, Ulan-Usu river valley 7–8 km beyond its emergence from mountains, grassy spruce grove, July 24, 1957—Yun. et al.).

IIIA. Qinghai: *Amdo* (along Mudzhik river, 2700 m, in humid scrubs, rare, June 16, 1880—Przew.).

General distribution: Jung.-Tarb., Nor. Tien Shan; Mid. Asia (Pam.-Alay), West. Sib. (Altay), East. Sib. (Ang.-Sayan), Nor. Mong. (Hent.).

6. Epipactis Zinn.
Cat. pl. (1757) 85.

1.   Labellum 11–13 mm long, its rear portion (hypochile) slightly concave, with obtuse auricles on sides and set off from flat

broadly oval front portion (epichile) by narrow fold. Rhizome creeping, with long internodes; leaves 4–6 ....
.... 2. **E. palustris** (L.) Crantz.
+ Labellum 7–9 mm long, its rear portion concave dish-like, without auricles, smaller than ovate-cordate front portion. Rhizome short, creeping .... 2.
2. Caulous leaves 6–10 .... 1. **E. helleborine** (L.) Crantz.
+ Caulous leaves 2 (rarely 1–3) .... 3. **E. tangutica** Schlechter.

1. **E. helleborine** (L.) Crantz, Stirp. Austr. 2 (1769) 467, p.p.; Mansfeld in Feddes repert. 45 (1938) 238. —*E. latifolia* (L.) All. Fl. Pedem. 2 (1785) 151; Kryl. Fl. Zap. Sib. 3 (1929) 703; Kraenzl. in Feddes repert., Beih. 65 (1931) 74; Nevski in Fl. SSSR, 4 (1935) 624; Fl. Kirgiz. 3 (1951) 136; Fl. Kazakhst. 2 (1958) 261. —*Serapias helleborine* α. *latifolia* L. Sp. pl. (1753) 949. —**Ic.:** Reichb. Ic. Fl. Germ. 13–14, tab. 136; Fl. SSSR, 4, Plate XXXIX, fig. 3.
Described from Europe. Type in London (Linn.).
Forest and coastal meadows, spruce groves in forest belt.

IIA. Junggar: *Tien Shan* (Pilyuchi gorge near Kul'dzha, June 1877; south. bank of Sairam lake, March 20, 1877; Talki gorge, 1800–2400 m, Aug. 16, 1877; Borgaty brook, nor. side of Kash river valley, 1500–1800 m, July 5, 1879; Aryslyn [Kash river valley], 2400 m, July 17, 1879; Aryslyn near estuary, about 1700 m, July 22, 1879—A. Reg.).

General distribution: Jung.-Tarb., Nor. and Cent. Tien Shan; Europe, Mediterr. (west.), Balk.-Asia Minor, Fore Asia (nor.), Caucasus, West. Sib., East. Sib. (Ang.-Sayan.).

2. **E. palustris** (L.) Crantz, Stirp. Austr. 2 (1769) 462; Kryl. Fl. Zap. Sib. 3 (1929) 702; Kraenzl. in Feddes repert., Beih. 65 (1931) 72; Nevski in Fl. SSSR, 4 (1935) 623; Fl. Kirgiz. 3 (1951) 136; Fl. Kazakhst. 2 (1958) 261; Fl. Tadzh. 2 (1963) 404. —*Serapias helleborine* η. *palustris* L. Sp. pl. (1753) 950. —**Ic.:** Reichb. Ic. Fl. Germ. 13–14, tab. 131; Fl. SSSR, 4, Plate XXXIX, fig. 5.
Described from Europe. Type in London (Linn.).
Humid and marshy meadows, in tugais, banks of irrigation ditches.

IIA. Junggar: *Tien Shan* (Yuldus-Taldy, 3000–3350 m, May 26; Nilki foothills, 1500 m, June—1879, A. Reg.); *Zaisan* (Kaba river around Kaba village, tugai, June 16; between Kara area and Kaba village, along bank of irrigation ditch, June 16—1914, Schischk.); *Dzhark.* (around Kul'dzha [along irrigation ditches], 1876—Golike; Suidun, July 1879—A. Reg.).

General distribution: Jung.-Tarb., Nor. Tien Shan; Europe, Balk.-Asia Minor, Caucasus, Mid. Asia (West. Tien Shan), West. Sib., East. Sib. (Irkutsk region).

3. **E. tangutica** Schlechter in Feddes repert., Beih. 4 (1919) 57, 149; id. in Acta Horti Gothoburg. 1 (1924) 145.
Described from Qinghai. Type in Wroclaw (WRSL)? Isotype in Leningrad.
Forest and coastal meadows, banks of brooks and irrigation ditches.

136

IIIA. Qinghai: *Nanshan* (Kha-Gomi in Huang He upper course, 2100 m, along irrigation ditches, on silty soil, common, July 4; South Tetung mountain range, forest, Aug. 7—1880, Przew., typus!; Lovachen town, fringe of irrigation ditch on loamy soil, July 26, 1908—Czet.).

General distribution: China (South-west: nor.-west. Sichuan).

7. Spiranthes Rich.

Orch. Europ. Annot. (1817) 20, 28, 36; id. in Mém.
Mus. nat. hist. natur. (Paris) 4 (1818) 50.

1. Labellum subsessile, oblong-ovate, without constriction; perianth 5–7 mm long; inflorescence 8–12 mm broad ....
.... 1. S. amoena (M.B.) Spreng.

+ Labellum with short claw, with distinct constriction at middle, panduriform; perianth usually 1.5 times smaller and inflorescence correspondingly narrower ....
.... 2. S. sinensis (Pers.) Ames.

1. **S. amoena** (M.B.) Spreng. Syst. 3 (1826) 708; Schlechter in Feddes repert., Beih. 4 (1919) 160; Nevski in Fl. SSSR, 4 (1935) 638; Kitag. Lin. Fl. Mansh. (1939) 155; Grub. Konsp. fl. MNR (1955) 98; Fl. Kazakhst. 2 (1958) 262; Kryl. Fl. Zap. Sib. 12 (1961) 3216. —*S. australis* Lindl. Bot. Regist. (1824) tab. 823; ej. Gen. a Sp. Orchid. (1840) 464, p.p., non *Neottia australis* R. Br.; Forbes and Hemsley, Index Fl. Sin. 3 (1903) 41, p.p.; Kryl. Fl. Zap. Sib. 3 (1929) 706; Kraenzl. in Feddes repert., Beih. 65 (1931) 83, p.p. — *Neottia amoena* M.B. Fl. taur.-cauc. 3 (1819) 606. —Ic.: Fl. SSSR, 4, Plate XXXIX, fig. 8.

Described from Siberia. Type in Leningrad.

Coastal solonetz meadows.

IA. **Mongolia:** *Cent. Khalkha* (Kerulen river valley near Bain-Erkhetu hill, 1889—Pal.; Toly river valley, Ulkhuin-Bulun area, loam and solonchak, Aug. 10, 1925—Gus.); *East. Mong.* (Khailar town, meadow and pasture steppe, 1959—Ivan.; right bank of Huang He river below Hekou town, sandy-clayey soil, Aug. 4, 1884—Pot.); *Depr. Lakes* (Kharkhiry river canal 3–4 km south of Ulangom, wet hummocky solonchak-like meadow, July 28, 1945—Yun.); *Ordos* (Huang He valley [in Tsaidamin-Nor lake region], in flooded meadow, on humus soil, common, Aug. 9, 1871—Przew.; 25 km south-east of Otokachi town, Kaolaitunao lake, solonchak meadow, Aug. 1; 10 km south-west of Ushinchi town, in meadow among willow and sea buckthorn thickets, Aug. 4—1957, Petr.).

IIA. **Junggar:** *Zaisan* (Kaba river around Kaba village, tugai, June 16, 1914—Schischk.).

General distribution: Europe (nor.-east European USSR), West. Sib., East. Sib., Far East (south.), Nor. Mong. (Hent., Hang.?, Mong.-Daur.), China (Dunbei), Korean peninsula.

2. **S. sinensis** (Pers.) Ames. Orchid. 2 (1908) 53; Schlechter in Feddes repert., Beih. 4 (1919) 160; id. in Acta Horti Gothoburg. 1 (1924) 147;

Nevski in Fl. SSSR, 4 (1935) 639; Walker in Contribs. U.S. Nat. Herb. 28 (1941) 605. —*S. australis* Lindl. Bot. Regist. (1824) tab. 823; ej. Gen. a. sp. Orchid. (1840) 464, p.p.; ?Franch. Pl. David. 1 (1884) 295; Forbes and Hemsley, Index Fl. Sin. 3 (1903) 41, p. max. p.; Kraenzl. in Feddes repert., Beih. 65 (1931) 83, p.p. —*Neottia sinensis* Pers. Syn. 2 (1807) 511. —**Ic.:** Fl. SSSR, 4 Plate XXXIX, fig. 9.

Described from South. China (Canton district). Type in Leiden (L.).
Wet coastal and forest meadows.

IIIA. Qinghai: *Nanshan* (South Tetung mountain range, coniferous forest, Aug. 8; between Choibsen temple and South Tetung mountain range, on hummocks, July 29—1880, Przew.; Tanfanza village [Tetung river valley], July 23; Paba-Tsyuan' valley, July 23; Shibanguku area, wet meadow among irises, common, Aug. 4—1908, Czet.; "Shang Hsin Chuang, scattered on exposed moist grassland, common, No. 679, 1923—R.C. Ching"—Walker, l.c.).

General distribution: Far East (south.), China (excluding Dunbei), Korean peninsula, Japan, Indo-Mal.

8. **Goodyera** R. Br.
in Alt. Hort. Kew., ed. 2, 5 (1813) 197.

1. **G. repens** (L.) R. Br. l.c. 198; Hook. f. Fl. Brit. Ind. 6 (1894) 111; Forbes and Hemsley, Index Fl. Sin. 3 (1903) 45; Schlechter in Feddes repert., Beih. 4 (1919) 166; Kryl. Fl. Zap. Sib. 3 (1929) 712; Kraenzl. in Feddes repert., Beih. 65 (1931) 84; Nevski in Fl. SSSR, 4 (1935) 639; Kitag. Lin. Fl. Mansh. (1939) 152; Grub. Konsp. fl. MNR (1955) 98; Fl. Kazakhst. 2 (1958) 263. —*Satyrium repens* L. Sp. pl. (1753) 945. —**Ic.:** Reichb. Ic. Fl. Germ. 13–14, tab. 130; Fl. Kazakhst. 2, Plate XXIV, fig. 9.

Described from Europe (Sweden). Type in London (Linn.).

Shady moss-covered coniferous and deciduous forests (with grassy cover), shaded moist rocks in forest belt, 1800–2400 m alt.

IIA. Junggar: *Tien Shan* (east up to Bogdo-Ula).
IIIA. Qinghai: *Nanshan* (South Tetung mountain range, about 2300 m, deciduous forest, among mosses, common, Aug. 10, 1880—Prezw.; Chorten-Ton temple, 2100 m, spruce forest, on humus and mosses, common and in clusters, Sept. 1–10, 1901—Lad.).

General distribution: Jung.-Tarb., Nor. Tien Shan; Europe, Balk.-Asia Minor, Fore Asia, Caucasus, West. and East. Sib., Far East, Nor. Mong., China (Dunbei, Cent.), Himalayas (east.), Korean peninsula, Japan, Nor. Amer.

9. **Habenaria** Willd.
Sp. pl. 4 (1805) 44.

1. **H. spiranthiformis** Ames et Schlechter in Feddes repert., Beih. 4 (1919) 52, 134; Schweinf. in J. Arn. Arb. 10 (1929) 172; Rehder in J. Arn. Arb. 14 (1933) 8.

Described from China (Yunnan). Type in Berlin (B) ?
Coastal meadows.

IIIA. Qinghai: *Amdo* ("Radja and Yellow river gorges, grassy sand-banks of Yellow river, south-east of Radja, alt. 3000 m, No. 14210, June 1926 [flowers green], J. Rock"—Schweinf. l.c.).
General distribution: China (South-west).

Note. Neither type nor authentic specimens of this species were studied.

10. Herminium Guett.
in Mém. Hist. Ac. Sci. Paris 1750 (1754) 374.

1. Leaves 2–4, sessile. Labellum 3-lobed; bracts lanceolate, long ..... 2.
+ Leaf 1, petiolate. Labellum undivided; bracts orbicular-deltoid, very short .... 3.
2. Leaves linear or linear-lanceolate, 2–4. Labellum with saccate spur and 3 nearly equal lobes .... 1. **H. alaschanicum** Maxim.
+ Leaves broadly lanceolate or elliptical. Labellum without spur, its lateral lobes considerably shorter than middle lobe ....
.... 2. **H. monorchis** (L.) R. Br.
3. Base of labellum with 2 emarginate pores along sides of midnerve; inflorescence with distant flowers on long pedicels, longer than bracts .... 3. **H. biporosum** Maxim.
+ Labellum without pores, inflorescence compact, with subsessile flowers ....4. **H. pugioniforme** Lindl.

1. **H. alaschanicum** Maxim. in Bull. Ac. Sci. St.-Pétersb. 31 (1886) 105, incl. var. *alaschanicum* Maxim. and var. *tanguticum* Maxim. ibid. —*H. tanguticum* Rolfe in J. Linn. Soc. (London) Bot. 36 (1903) 51 (in Index Fl. Sin.); Schlechter in Feddes repert., Beih. 4 (1919) 103; Schweinf. in J. Arn. Arb. 10 (1929) 171; Walker in Contribs. U.S. Nat. Herb. 28 (1941) 605. —*H. altigenum* Schlechter in Feddes repert., Beih. 12 (1922) 334; id. in Acta Horti Gothoburg. 1 (1924) 135.

Described from Mongolia (Alashan) and Qinghai (Nanshan). Type in Leningrad. Plate V, fig. 4.

Coniferous forests, scrubs, alpine meadows, moss-covered rocks, up to 4700 m alt.

IA. Mongolia: *Alash. Gobi* (Alashan mountain range, on exposed clayey slope, occasional, No. 163, June 27 [July 9] 1873—Przew., syntypus!).

IIIA. Qinghai: *Nanshan* (alpine belt between Nanshan mountain range and Donkyru area on Rako-Gol river, 3000–3350 m, on rocks, rare, No. 535, July 10 [22], 1880—Przew., syntypus!; "Li Fu Yai, on moist rocky cliff, very rare, No. 462, 1923, R.C. Ching"—Walker, l.c.).

IIIB. Tibet: *Weitzan* (Yangtze river basin, on Khichu river, 3700 m, Aug. 2, 1900—Lad.).

General distribution: China (North-west, South-west).

Note. A study of the authentic specimens of this species revealed teratological abnormalities in the characteristics of the form of perianth lobes in Tangut plants described by K.I. Maximowicz not only in a given plant specimen, but even in a single flower. These abnormalities were also observed in some flowers of *Herminium monorchis* (L.) R. Br. For this reason, K.I. Maximowicz' division of his new species into 2 varieties—var. *alaschanicum* and var. *tanguticum* Rolfe—is irrational. Schlechter (op. cit.) clearly adopted one of the forms of *H. monorchis* (L.) R. Br. differing in much longer lateral lobes of labellum as *H. alaschanicum* Maxim. and later described it as a new species, *H. altigenum* Schlechter.

2. **H. monorchis** (L.) R. Br. in Ait. Hort. Kew. 5 (1813) 191; Franch. Pl. David, 1 (1884) 295; Forbes and Hemsley, Index Fl. Sin. 3 (1903) 51; Danguy in Bull. Mus. nat. hist. natur. 17 (1911) 557; Schlechter in Feddes repert., Beih. 4 (1919) 102; ibid. Beih. 12 (1922) 333; id. in Acta Horti Gothoburg. 1 (1924) 134; Kryl. Fl. Zap. Sib. 3 (1929) 690; Kraenzl. in Feddes repert., Beih. 65 (1931) 62; Nevski in Fl. SSSR, 4 (1935) 643; Kitag. Lin. Fl. Mansh. (1939) 153; Grub. Konsp. fl. MNR (1955) 98; Fl. Kazakhst. 2 (1958) 263. —*Orchis monorchis* L. Sp. pl. (1753) 947. —**Ic.**: Reichb. Ic. Fl. Germ. 13–14, tab. 63; Fl. Kazakhst. 2, Plate XXV, fig. 1.

Described from Europe. Type in London (Linn.).

Moist, often solonetz coastal meadows, scrubs, forest meadows and glades.

IA. **Mongolia:** *Cent. Khalkha* ("Vallée de l'Orkhon, July 14, 1909—B. du Chazoud"—Danguy, l.c.); *East. Mong.* (Khailar town vicinity, in moist basis on sand, No. 1167, July 4, 1951—Lee et al.); Oulashan [Muni-Ula], liex frais, elevés, No. 2794, July 1866—David); *Depr. Lakes* (Ulangom, meadows, Sept. 5, 1879—Pot.).

IIIA. **Qinghai:** *Nanshan* (near Kuku-Nor lake, 3100 m, July 9; North Tetung mountain range, 2400 m, July 14; between Choibsen temple and South Tetung mountain range, 2600 m, on hummocks, July 29—1880, Przew.; Shibanguku area, along river-bank among poplars, common, Aug. 4, 1908—Czet.; 65 km south-east of Chzhan'e town, high Nanshan foothills, 2300 m, small meadows in mountain valley, July 12, 1958—Petr.); *Amdo* (Mudzhik river [south of Guidui], 2750 m, in moist scrubs on silty shoal, June 16, 1880—Przew.).

IIIB. **Tibet:** *Weitzan* (Darindo area near Chzherku monastery, Bounchin rocks, 3550 m, wet mosses and humus below rocks and among scrubs, common and in clusters, Aug. 9, 1900—Lad.); *South.* (Gyantse, July-Sept. 1904—Walton).

General distribution: Jung.-Tarb.; Europe, Mediterr. (nor.-west.), Balk., West. and East. Sib., Far East (south.), Nor. Mong., China (Dunbei, North, North-west, East, South-west), Himalayas.

Note. Depending on habitat conditions, specimens of the species vary greatly in height from 6 to 25 and even up to 40 cm also, labellum with lateral lobes much shorter than middle lobe and recurved outward or nearly equal to middle lobe and parallel to it.

3. **H. biporosum** Maxim. in Bull. Ac. Sci. St.-Pétersb. 31 (1886) 106; Forbes and Hemsley, Index Fl. Sin. 3 (1903) 51; Schlechter in Feddes

repert., Beih. 4 (1919) 100. —*Porolabium biporosum* (Maxim.) Tang et Wang in Bull. Fan. mem. Inst. Biol. 10 (1940) 38.

Described from Qinghai (Kuku-Nor). Type in Leningrad.

IIIA. Qinghai: *Nanshan* (Viciniis Kuku-Nor, medio Julio 1880—Prz., typus!).
General distribution: China (North: Shanxi; North-west?).

Note. Only a single type specimen of this species, closely related to the Himalayan *H. gramineum* Lindl., is known to us. It is a small plant, about 8 cm tall. Labellum pores are also known in some genetically closely related Himalayan species but the labellum and column different somewhat in shape and thus the classification of genus *Porolabium* Tang et Wang on the basis of this characteristic is, in our opinion, unfounded.

4. H. pugioniforme Lindl. ex Hook. f. Fl. Brit. India, 6 (1894) 130. —*H. nivale* Schlechter in Acta Horti Gothoburg. 1 (1924) 134.

Described from Himalayas. Type in London (K).

Alpine meadows, cracks and crevices in rocks, up to 5000 m alt.

IIIB. Tibet: *Weitzan* (Ichu river upper course, Yantszytszyan tributary, 3950 m, in rock crevices on humus, July 29, 1900—Lad.).
General distribution: China (South-west), Himalayas.

11. Neottianthe Schlechter
in Feddes repert. 16 (1919) 290.

1. N. cucullata (L.) Schlechter in Feddes repert. 16 (1919) 292; id. in Feddes repert., Beih. 12 (1922) 333; id. in Acta Horti Gothoburg. 1 (1924) 137; Nevski in Fl. SSSR, 4 (1935) 645; Kitag. Lin. Fl. Mansh. (1939) 153; Grub. Konsp. Fl. MNR (1955) 98; Fl. Kazakhst. 2 (1958) 264. —*Orchis cucullata* L. Sp. pl. (1753) 939. —*Gymnadenia cucullata* Rich. in Mém. Mus. hist. natur. (Paris) 4 (1818) 57; Forbes and Hemsley, Index Fl. Sin. 3 (1903) 52; Danguy in Bull. Mus. nat. hist. natur. 17 (1911) 557; Schlechter in Feddes repert., Beih. 4 (1919) 105; Kryl. Fl. Zap. Sib. 3 (1929) 695; Kraenzl. in Feddes repert., Beih. 65 (1931) 49. —*Habenaria cucullata* Hoefft, Cat. Pl. Koursk (1826) 56; Schweinf. in J. Arn. Arb. 10 (1929) 171; Walker in Contribs. U.S. Nat. Herb. 28 (1941) 604. —Ic.: Reichb. Ic. Fl. Germ. 13–14, tab. 66; Fl. Kazakhst. 2, Plate XXV, fig. 2.

Described from Siberia. Type in London (Linn.).

Moist moss-covered coniferous forests.

IIIA. Qinghai: *Nanshan* (South Tetung mountain range, 2300 m, coniferous forest, on humus soil, common; sand-bank of Tetung river, common and in clusters—Aug. 8, 1880, Przew.).
General distribution: Jung.-Tarb. (Jung. Alatau); Europe, West. Sib., East. Sib. (Sayans, Daur.), Far East (south.), Nor. Mong. (Hent.), China (Dunbei, North, North-west, Cent., South-west), Korean peninsula, Japan.

12. **Coeloglossum** C. Hartm.
Handb. Scand. Fl. (1820) 329.

1. **C. viride** (L.) C. Hartm. l.c.; Kryl. Fl. Zap. Sib. 3 (1929) 691; Nevski in Fl. SSSR, 4 (1935) 647; Fl. Kirgiz. 3 (1951) 142; Grub. Konsp. fl. MNR (1955) 98; Fl. Kazakhst. 2 (1958) 265; Fl. Tadzh. 2 (1963) 406. —*C. bracteatum* Parl. Fl. Ital. 3 (1850) 409; Schlechter in Feddes repert., Beih. 12 (1922) 335; id. in Acta Horti Gothoburg. 1 (1924) 137. —*Satyrium viride* L. Sp. pl. (1753) 944. —*Orchis bracteata* Willd. Sp. pl. 4 (1805) 34. —*Peristylus viridis* Lindl. Syn. Brit. Fl. (1835) 261; Forbes and Hemsley, Index Fl. Sin. 3 (1903) 54. —*Platanthera viridis* Lindl. Syn. Brit. Fl. (1829) 261; Kraenzl. in Feddes repert., Beih. 65 (1931) 52. —*P. viridis* Lindl. var. *bracteata* Reichb. f. Ic. Fl. Germ. 13–14 (1951) 164. —**Ic.:** Reichb. l.c., tab. 82.

Described from Nor. Europe. Type in London (Linn.).

Coniferous sparse forests and their fringes, scrubs, forest and subalpine meadows, turf-covered rock placers and rocks, 1800–3500 m alt.

IIA. Junggar: *Tarb.* (south. slope of Saur mountain range, Karagaitu river valley, creek valley on right bank of Bain-Tsagan, subalpine meadow beyond larch forest, June 23, 1957—Yun. et al.); *Tien Shan* (Ili river basin).

IIIA. Qinghai: *Nanshan* (South Tetung mountain range, about 2300 m, coniferous forest, rare, Aug. 9, 1880—Przew.; South Kukunor mountain range, south. bank of Kuku-Nor, on nor. slope, 3100–3500 m, rare, Aug. 20, 1901—Lad.); *Amdo* (Mudzhik mountains, alpine zone, humid forest, common, June 22, 1880—Przew.).

IIIB. Tibet: *Weitzan* (Yangtze river, Nruchu area, about 3550 m, below rocks shaded by overgrowth of juniper forest, July 25; Yangtze river-basin, near Chzherku temple, Bounchin rocks in Darindo area, about 3550 m, on wet humus, Aug. 9—1900, Lad.).

General distribution: Jung.-Tarb., Nor. Tien Shan, Cent. Tien Shan; Europe, Balk.-Asia Minor, Caucasus, Mid. Asia (West. Tien Shan, Alay), West. and East. Sib., Far East, Nor. Mong., China (Nor., Cent.), Korean peninsula, Japan, Nor. Amer. (nor.).

Note. The form with highly developed, long bracts [var. *bracteatum* (Willd.) Reichb. f.] is found at various places within the vast distribution range of this species, predominantly in mountains, from Europe to Japan and Nor. America, along with the type form. The extent of bract development also varies markedly and it is hardly possible to draw any objective border between type form and form with long bracts. We therefore subscribe to S.A. Nevski's (l.c.) view that the latter should not be treated [as well as the dwarfish alpine form var. *islandicum* (Lindl.) M. Schulze] as a distinct species and regard the genus as monotypic.

13. **Platanthera** Rich.
Orch. Europ. Annot. (1817) 20, 26, 35; id. in Mém. Mus.
hist. natur. (Paris) 4 (1818) 42, 48, 51.

1. **P. freynii** Kraenzl. in J. Bot. Russ. (1913) 37; id. in Feddes repert., Beih. 65 (1931) 58; Nevski in Fl. SSSR, 4 (1935) 659; Kitag. Lin. Fl. Mansh.

(1939) 154. —*P. densa* Freyn in Oesterr. Bot. Zeitschr. 46 (1896) 96, non Lindl. (1828). —*P. chlorantha* Cust. var. *orientalis* Schlechter in Feddes repert., Beih. 4 (1919) 109; ibid. 12 (1922) 337. —*P. chlorantha* auct. non Cust.: Forbes and Hemsley, Index Fl. Sin. 3 (1903) 55; Schlechter in Acta Horti Gothoburg. 1 (1924) 137. —*Habenaria bifolia* auct. non R. Br.: Walker in Contribs. U.S. Nat. Herb. 28 (1941) 604. —**Ic.:** Fl. SSSR, 4, Plate XL, fig. 11.

Described from East. Siberia (Dauria, Nerchinsk environs), type in Praga [Prague] (PR).

Coniferous, deciduous and mixed forests, scrubs, forest and alpine meadows.

IIIA. Qinghai: *Nanshan* (South Tetung mountain range, mixed forest, common and sometimes in clusters, July 11, 1872—Przew.; "Tu Er Ping, in a dense *Picea* forest, No. 349; Ching Kang Yai, in woods, rare, No. 570 [1925—R.C. Ching], height up to 60 cm, flowers creamy-white or yellowish-green, anthers orange-yellow"—Walker, l.c.).

General distribution: East. Sib. (Daur. east.), Far East (south), China (Dunbei, North, North-west, South-west), Korean peninsula, Japan.

14. Lysiella Rydb.
in Mém. New York Bot. Gard. 1 (1900) 104.

1. **L. oligantha** (Turcz.) Nevski in Fl. SSSR, 4 (1935) 663; Grub. in Bot. zh. 56, 11 (1971) 1642. —*Platanthera oligantha* Turcz. in Bull. Soc. Natur. Moscou, 27 (1854) [Fl. baic.-Dahur.] 86. —*P. parvula* Schlechter in Feddes repert. 15 (1918) 301; id. in Feddes repert., Sonderbeih. A. (1928) 253. —**Ic.:** Schlechter, l.c. (1928) tab. 31, fig. 121; Fl. SSSR, 4, Plate XL, fig. 13.

Described from East. Siberia (Baikal territory). Type in Kiev (KW), isotype in Leningrad.

Shady coniferous forests.

IIA. Junggar: *Tien Shan* (Bai town region, in forest belt, No. 2031, Sept. 22, 1957—Shen'-Tyan').

General distribution: Europe (Scand.), East. Sib., Far East (nor.), Nor. Mong. (Fore Hubs., Hang.).

Note. The cited locality of this species is unexpected, isolated and very far from the main distribution range in East. Siberia. The specimen was, in all probability, collected in a forest of spruce *Picea schrenkiana* Fisch. et Mey. on the south. slope of Tien Shan.

15. Gymnadenia R. Br.
in Ait. Hort. Kew., ed. 2, 5 (1813) 191.

1. **G. conopsea** (L.) R. Br. in Ait. Hort. Kew., ed. 2, 5 (1813) 191; Forbes and Hemsley, Index Fl. Sin. 3 (1903) 52; Schlechter in Feddes repert., Beih. 4 (1919) 104; ibid. 12 (1922) 332; Kryl. Fl. Zap. Sib. 3 (1929) 693; Kraenzl. in Feddes repert., Beih. 65 (1931) 49; Nevski in Fl. SSSR, 4 (1935) 668;

Kitag. Lin. Fl. Mansh. (1939) 152; Grub. Konsp. fl. MNR (1955) 98; Fl. Kazakhst. 2 (1958) 266. —*Orchis conopsea* L. Sp. pl. (1753) 942. —*Habenaria conopsea* Benth. in J. Linn. Soc. (London) Bot. 18 (1881) 354; Schweinf. in J. Arn. Arb. 10 (1929) 171; Walker in Contribs. U.S. Nat. Herb. 28 (1941) 604. —**Ic.**: Reichb. Ic. Fl. Germ. 13–14, tab. 70–73; Fl. SSSR, 4, Plate XL, fig. 14. Described from Europe. Type in London (Linn.).

Coniferous and deciduous forests, forest meadows and fringes, scrubs.

IA. Mongolia: *East. Mong.* (Inshan': Muni-Ula [July 6–18] 1871—Przew.).

General distribution: Europe, Mediterr., Balk.-Asia Minor, Fore Asia, Caucasus, West. Sib., East. Sib., Nor. Mong., China (Dunbei, North, North-west, East, South-west), Korean peninsula, Japan.

16. Galeorchis Rydb.
in Britton, Manual (1901) 292.

1. Flowers white; labellum without spur, almost indistinguishable in shape and size from rest of lobes of inner whorl of perianth. Leaf solitary, radical, elliptical; sometimes one small caulous, lanceolate leaf also developed ....
.... 1. **G. albiflora** (Schlechter) Grub.
+ Flowers pink; labellum with spur. Leaf solitary .... 2.
2. Labellum undivided; spur much shorter than ovary ....
.... 2. **G. reichenbachii** Nevski.
+ Labellum 3-lobed; spur as long as ovary ....
.... 3. **G. roborovskii** (Maxim.) Nevski.

1. **G. albiflora** (Schlechter) Grub. comb. nova. —*Aceratorchis albiflora* Schlechter in Feddes repert., Beih. 12 (1922) 328. —*A. tschiliensis* auct. non Schlechter (1922): (?) Schlechter in Acta Horti Gothoburg. 1 (1924) 170; Schweinf. in J. Arn. Arb. 10 (1929) 170.

Described from China (Yunnan). Type in Wroclaw (WRSL) ?

Alpine meadows.

IIIA. Qinghai: *Amdo* ("Radja and Yellow river gorges: alpine meadows, mountains south-west of Radja; south of river, alt. 3600–3900 m, No. 14190, June 1926 [flowers white]"—Schweinf. l.c.).

IIIB. Tibet: *Weitzan* (Yangtze river basin, Darindo area near Chzherku temple, Bounchin rocks, 3550 m, on rocks, moist humus, common, Aug. 9, 1900—Lad.).

General distribution: China (South-west).

Note. Apart from the absence of spur, this species, like the closely related pink-coloured *G. tschiliensis* (Schlechter) Grub. comb. nova (*Aceratorchis tschiliensis* Schlechter) from North. China, differs in no other significant feature from the rest of the spur-bearing species of this genus. Schlechter (l.c.) and Schweinfurth (l.c.) suggested the probable existence of simple peleoric forms corresponding to spur-bearing species but this view is contradicted by their frequent occurrence and definite geographic affinity (distribution range).

2. **G. reichenbachii** Nevski in Fl. URSS, 4 (1935) 670. —*Orchis spathulata* Reichb. f. ex Benth. in J. Linn. Soc. (London) Bot. 18 (1881) 355, non L. f. (1781); Hook. f. Fl. Brit. India, 6 (1894) 127; Forbes and Hemsley, Index Fl. Sin. 3 (1903) 50; Schlechter in Feddes repert., Beih. 4 (1919) 91; ?Schweinf. in J. Am. Arb. 10 (1929) 170. —? *O. spathulata* Reichb. f. var. *wilsonii* Schlechter in Acta Horti Gothoburg. 1 (1924) 132. —*Gymnadenia spathulata* Lindl. Gen. a. Sp. Orchid. (1835) 280. —**Ic.:** Hook. Ic. Pl. tab. 2197, fig. A.

Described from Himalayas. Type in Leiden (L).

Coniferous forests, forest glades and fringes, alpine meadows.

IIIA. **Qinghai:** *Nanshan* (Chortyn-Ton temple [on Tetung river], in spruce forest, on humus, No. 593, Sept. 1–10, 1901—Lad.).

General distribution: China (South-west), Himalayas.

3. **G. roborovskii** (Maxim.) Nevski in Fl. URSS, 4 (1935) 670 in nota. —*Orchis roborovskii* Maxim. in Bull. Ac. Sci. St.-Pétersb. 31 (1886) 104; Forbes and Hemsley, Index Fl. Sin. 3 (1903) 50; Schlechter in Feddes repert., Beih. 4 (1919) 90. —*O. szechenyiana* Reichb. f. in Kanitz, Pl. Exped. Széchenyi Asia centr. enumer. (1891) 58; id., Die Resultate der botanischen Sammlungen...(1898) 732; Forbes and Hemsley, Index Fl. Sin. 3 (1903) 50; Schlechter in Feddes repert., Beih. 4 (1919) 91; id. in Acta Horti Gothoburg. 1 (1924) 132. —**Ic.:** Kanitz, l.c. (1898), tab. VI, fig. I, 1–5.

Described from Qinghai (Nanshan). Type in Leningrad.

Alpine meadows, scrubs, rocks.

IIIA. **Qinghai:** *Nanshan* (along Yusun-Khatyma river, 3050 m, scrubs on rocks, rare, No. 562, July 12 [24] 1880—Przew., typus!; between Nanshan and Donkyru, in alpine belt, 3050–3350 m, on rocks, quite rare, July 22, 1880—Przew.; "Altin-Gomba in ditione Sining-fu, 2780 m s.m., July 7, 1879, Szécheny"—Kanitz, l.c. [1898]).

General distribution: China (South-west: nor. Sichuan).

Note. While describing this species, K.I. Maximowicz (l.c.) pointed to the presence of 2 leaves, counting the lower much larger leaf-like bract. In fact, this species has only a solitary normally developed radical leaf. He cited the number of flowers in the inflorescence as 2–3 but actually this number varies in a very wide range of 1–8. These inaccuracies of description were the cause of *Orchis szechenyiana* Reichb. f. being treated as a distinct species by later investigators although it does not differ at all from *G. roborovskii* except for the large number of flowers in the inflorescence. A comparison of authentic specimens of *O. roborovskii* Maxim. (and the author's drawing of flower appended to it) with the drawing of *O. szechenyiana* Reichb. f. (l.c.) shows that they pertain to the same species.

17. **Ponerorchis** Reichb. f.

in Linnaea, 25 (1852) 227. —Chusua Nevski in Fl. SSSR, 4 (1935) 753, 670.

1. **P. pauciflora** (Lindl.) Ohwi in Acta phytotax. geobot. (Kyoto) 5 (1936) 145; Kitag. Lin. Fl. Mansh. (1939) 155 ("*Poneorchis*"). —*Gymnadenia pauciflora* Lindl. Gen. a. Sp. Orchid. (1835) 280. —*Orchis pauciflora* Fisch.

ex Lindl. l.c. 280, pro syn. non Tenore, 1811; Schlechter in Feddes repert., Beih. 4 (1919) 90; Kraenzl. in Feddes repert., Beih. 65 (1931) 32. —*O. chusua* auct. non D. Don: Walker in Contribs. U.S. Nat. Herb. 28 (1941) 605. — *Chusua secunda* Nevski in Fl. SSSR, 4 (1935) 670. —**Ic.:** Reichb. Ic. Fl. Germ. 13–14, tab. 170, fig. 1; Fl. SSSR, 4, Plate XLII, fig. 10.

Described from East. Sib. (Dauria). Type in Leningrad.

Shady moss-covered coniferous forests, montane and forest meadows, moist shaded rocks.

IIIA. **Qinghai:** *Nanshan* (Yusun-Khatyma river, 3050 m, on rocks among scrubs, common, July 24, 1880—Przew.; "Tu Er Ping, No. 428; Li Fu Yai, No. 473, in forests, common [1923, R.C. Ching]; height up to 40 cm; flowers purplish, dotted inside with deeper-coloured spots"—Walker, l.c.).

General distribution: East. Sib. (Ang.-Sayan, Daur.), Far East, (south.), China (Dunbei).

18. Orchis L.
Sp. pl. (1753) 939.

1.　Leaves without spots, linear-lanceolate, upper ones reaching base of inflorescence or even longer, large and differing sharply from bracts. Labellum undivided .... 2.

+　Leaves with dark spots above; lower ones oblong-obovate, obtuse; upper small, bract-like and considerably away from base of inflorescence. Labellum 3-lobed .... 1. **O. fuchsii** Druce.

2.　Leaves strongly declinate from stem; lower ones arcuately curved and usually longitudinally folded; stem 15–20 cm tall. Bulbs 3–5-partite. Labellum 7–9 mm long and as broad; spur nearly straight, 9–11 mm long .... 3. **O. salina** Turcz.

+　Leaves erect, poorly declinate from stem, flat; stem 25–50 cm tall. Flowers considerably smaller or larger; spur generally bent ...... 3.

3.　Leaf tip obtuse and constricted like a small hood. Flowers small; labellum 4.5–6 mm long and 5–7 mm broad; spur 5–6 mm long. Bulbs 2–4-partite .... 2. **O. latifolia** L.

+　Leaves acute or acuminate, without small hood. Flowers large; labellum 8–10 (12) mm long and 8–12 (14) mm broad; spur 12–15 (18) mm long. Bulbs 3–6-partite .... 4. **O. umbrosa** Kar. et Kir.

1. **O. fuchsii** Druce, Rep. Bot. Exch. Cl. Brit. Isles, 4 (1914) 105; Nevski in Fl. SSSR, 4 (1935) 704; Grub. Konsp. fl. MNR (1955) 98; Fl. Kazakhst. 2 (1958) 270; Kryl. Fl. Zap. Sib. 12 (1961) 3210. —*O. maculata* auct. non L.: Sap. Mong. Altay (1911) 387; Kryl. Fl. Zap. Sib. 3 (1929) 688; Kraenzl. in Feddes repert., Beih. 65 (1931) 42, p.p. —**Ic.:** Fl. SSSR, 4, Plate XLII, fig. 1.

Described from Europe (British Isles). Type in London (K).

Forest meadows, glades and fringes, scrubs, coastal meadows.

IIA. Junggar: *Cis-Alt.* (Mong. Altay, lower valley of Ulasta [tributary of Kran] river, forest glades, June 30, 1908—Sap.).

General distribution: Europe, West. and East. Sib., Nor. Mong. (Mong.-Daur.).

2. O. latifolia L. Sp. pl. (1753) 941, excl. var. *v*; ej. Fl. suec. (1755) 312; Danguy in Bull. Mus. nat. hist. natur. 20 (1914) 136; Kraenzl. in Feddes repert., Beih. 65 (1931) 38, p.p.; Nevski in Fl. SSSR, 4 (1935) 717; Fl. Kazakhst. 2 (1958) 271; Kryl. Fl. Zap. Sib. 12 (1961) 3208. —*O. incarnata* auct. non L.: Kryl. Fl. Zap. Sib. 3 (1929) 684; Kraenzl. l.c. 32, p.p. —Ic.: Reichb. Ic.: Fl. Germ. 13–14, tab. 45; Fl. SSSR, 4, Plate XLII, fig. 8.

Described from Europe. Type in London (Linn.).

Moist and marshy coastal meadows.

IA. Mongolia: *East. Mong.* (Kailar, alt. 750 m, terrains marecageux, June 22, 1896—Chaff.; Khaligakha area [south of Khuntu lake], on damp meadow, July 23, 1899—Pot. and Sold.).

IB. Kashgar: *South.* (Niya oasis, 1250 m, in marsh among rose shrubs, May 21-June 14, 1885—Przew.).

IIA. Junggar: *Jung. Alatau* (Toli district, Albakzin mountains, Koket, in marsh, No. 2677, Aug. 6, 1957—Kuan); *Tien Shan* (Kunges river, forest belt, on bank, July 24, 1877—Przew.; Talki, July 16; Dzhagastai, Aug. 7—1877; Bogdo mountain, 2100–2400 m, July; Koksu, 1500–1800 m, May 28—1878; Borgaty brook on north. bank of Kash river, 1500–1800 m, July 5; left bank of Kash river, 2750 m, July 15; between Saryk and Mengoto, 1800–2100 m, July 27; Aryslyn-Daban on right bank of Kash river, Aug. 7—1879, A. Reg.); *Zaisan* (between Kara area and Kaba village, bank of irrigation ditch, July 16, 1914—Schischk.).

General distribution: Jung.-Tarb.; Europe, Balk., West. Sib., East. Sib. (south.), Nor. Mong. (Hang.).

3. O. salina Turcz. ex Lindl. Gen. a. Sp. Orchid. (1835) 259; Forbes and Hemsley, Index Fl. Sin. 3 (1903) 50; Schlechter in Feddes repert., Beih. 4 (1919) 91; Schweinf. in J. Arn. Arb. 10 (1929) 170; Kryl. Fl. Zap. Sib. 3 (1929) 686; Kraenzl. in Feddes repert., Beih. 65 (1931) 35, p.p.; Nevski in Fl. SSSR, 4 (1935) 713; Kitag. Lin. Fl. Mansh. (1939) 154; Grub. Konsp. fl. MNR (1955) 99; Fl. Kazakhst. 2 (1958) 273. —*O. latifolia* auct. non L.: Danguy in Bull. Mus. nat. hist. natur. 17 (1911) 449, 556. —*O. latifolia* L. var. *salina* Trautv. in Tr. SPb. bot. sada, 1 (1872) 92; Sap. Mong. Altai (1911) 387. —Ic.: Reichb. Ic. Fl. Germ. 13–14, tab. 522.

Described from East. Siberia (Dauria). Type in Kiev (KW), isotype in Leningrad.

Solonetz moist and marshy meadows.

IA. Mongolia: *Mong. Alt.* (Tsitsirin-Gol gorge near estuary, in meadows, July 9, 1877—Pot.; upper Kobdo river 1 km below Khoton-Gol estuary, right bank, 2050 m, swampy sedge meadow, Aug. 11, 1971—Grub. et al.); *Khobd.* (Achit-Nur, nor.-west. bank, swampy Bukhu-Muren and Khub-Usu-gol interfluvial region 7–8 km from Bukhu-Muren somon, solonetz sedge meadow, July 15, 1971—Grub. et al.); *Cent. Khalkha* ("Entre Erden-Dzou and Karabalgassonne, July 5, 1909—Chazoud"—Danguy, l.c.; on bank of Kukshin-Orkhon, among shrubs and iris, July 17, 1893—Klem.; Toly meander, along Kharukha river valley near Ulan-Bumba area, solonetz meadow, July 14, 1924—Pavl.; Mongol-Elisu sand nor. of Ara-Dzhirgalantu-Gol, hummocky sand, July 3, 1949—Yun.); *East. Mong.* (Shabartai-Sume nor. of Dolon-Nor town, 1870—Lom.; near Kharkhonte railway station, basin between dunes, June 8, 1902—Litw.; Khailar town, saline low meadows, 1959—Ivan.); *Var. Lakes* (Tuni-Gol valley, meadow lowland, July 8, 1924—E. Gorbunova; Arguin-Gol, moist meadow, July 24; 1926—E. Kozlova); *Depr. Lakes* (left bank of Bukon-Beren' river on east. slope of Sailyugem, Altyn-Khatysyn area, June 17; Ulangom, in swamps, July 2—1879, Pot.).

IB. **Kashgar:** *Nor.* (Uch-Turfan, Kukurtuk gorge, on spring bank, May 27, 1908—Divn.); *West.* (Gumbus river, Chom-Terek area, in swamp, June 7, 1909—Divn.; Kizylsu basin, in river floodplain near Nagra-Galdy village, in grasslands, July 2, 1929—Pop.); *South.* (Keriyadar'ya river near Lyushi estuary, June 30, 1885—Przew.).

IIA. **Junggar:** *Tien Shan, Jung. Gobi* (south.: Dzhinkho river, 300 m, April 30; Sygashu, 550 m, May 13—1879, A. Reg.; Turkul' lake, along bank, June 16, 1877—Pot.; east.: floodplain of Bodonchi river 2–3 km before Bodonchin-Khure, brackish meadow, Aug. 1, 1947—Yun.; Uinchi lowland, Boro-Tsonchzhi area, solonetz sedge-wheat grass meadow, Sept. 13, 1948—Grub.); *Dzhark.* (west of Kul'dzha on Ili river-bank, May 17, 1877—A. Reg.).

IIIA. **Qinghai:** *Nanshan* (Nanshan alpine belt, July 1879; Kuku-Nor, July 14, 1880—Przew.; Humboldt mountain range, nor. slope, Kuku-Usu area, Blagodatnyi spring, 2700 m, in marsh, common and in clusters, May 28; same locality, 2700–3050 m, damp meadow, June 8—1894, Rob.; "Houei-Houei-Pou, 2000 m, June 19, 1908, Vaillant"—Danguy, l.c.; high foothills of Nanshan 65 km south-east of Chzhan'e town, 2300 m, small meadows in mountain valley, July 12, 1958—Petr.); *Amdo* (along Mudzhik river south of Guidui town, 2700 m, around springs among shrubs, common, June 16, 1880—Przew.).

IIIC. **Pamir** (Tashkurgan town, near irrigation ditch and on flooded meadow, July 25, 1903—Knorring; Tashkurgan river valley 1-2 km west of town, 3200 m, No. 320, June 13, 1959—Lee and Chu); same locality, 3–4 km west of town, along brooks and irrigation ditches, June 13, 1959—Yun. et al.).

General distribution: West. Sib. (Altay), East. Sib. (Ang.-Sayan., Daur.), Far East (Zee-Bur.), Nor. Mong., China (Dunbei).

4. **O. umbrosa** Kar. et Kir. in Bull. Soc. natur. Moscou, 15 (1842) 504; Nevski in Fl. SSSR, 4 (1935) 714; Fl. Kirgiz. 3 (1951) 141; Fl. Kazakhst. 2 (1958) 272; Ikonnik. Opred. rast. Pamira [Key to Plants of Pamir] (1963) 89; Fl. Tadzh. 2 (1963) 409. —*O. orientalis* Klinge ssp. *turkestanica* Klinge in Acta Horti Petrop. 17 (1895) 183. —*O. turkestanica* Klinge ex Fedtsch. in J. Bot. Russ. (1908) 191; Persson in Bot. notiser (1938) 278. —*O. incarnata* auct. non L.: Kraenzl. in Feddes repert., Beih. 65 (1931) 32, p.p. —*O. salina* auct. non Turcz.: Kraenzl. l.c. 35, p.p. —**Ic.:** Fl. SSSR, 4, Plate XLII, fig. 7.

Described from Kazakhstan (Semirech'e). Type in Moscow (MW) ?, isotype in Leningrad.

Humid coastal and valley meadows.

IB. **Kashgar:** *Nor.* (Bostan-Tograk village vicinity, 2500 m, near water, No. 9830, July 9, 1959—Lee and Chu); *West.* ("Bostan-Terek, about 2400 m, Aug. 8, 1933, Aug. 4, 1934—Persson).

IIA. **Junggar:** *Jung. Alatau* (im Ulan-Ussu Tal, Aug. 15–17, 1908—Merzb.); *Tien Shan* (Taldy river midcourse, 2100 m, May 26; Khanakhai brook south of Kul'dzha, 1200–1500 m, June 15—1879, A. Reg.; Khanga river valley 25 km beyond Balinte village along road to Yuldus from Karashar, meadow, Aug. 1, 1958—Yun. et al.); *Dzhark.* (Ili river left bank south of Kul'dzha, May 27; Khoir-Sumun south of Kul'dzha, May 27; Ili river-bank west of Kul'dzha, May 16—1877, A. Reg.); *Jung. Gobi* (west.: Tsagan-Usu, Dzhinkho branch, 975 m, June 8, 1879—A. Reg.).

IIIC. Pamir ("Tash-Korghan, Dafdar, about 3510 m, June 29, 1935; Yerzil, in damp meadow together with primula, about 3000 m, July 5, 1930"—Persson, l.c.; Issyksu river, 3000–3100 m alt., June 21; same locality, 3100 m, flooded meadow, July 3—1942, Serp.).

General distribution: Jung.-Tarb., Nor. and Cent. Tien Shan, East. Pam.; Mid. Asia (West. Tien Shan, Pam.-Alay), West. Sib. (south. fringe), Himalayas (west.).

Note. Difficult to distinguish from *O. salina* Turcz. because of several intermediate forms, probably of hybrid origin.

Plate I.
1—*Allium senescens* L.; 2—*A. bidentatum* Fisch. ex Prokh. (flower parts); 3—*A. mongolicum* Rgl.; 4—*A. prostratum* Trev. (flower); 5—*A. tenuissimum* L.; 6—*A. polyrhizum* Turcz. ex Rgl.

150

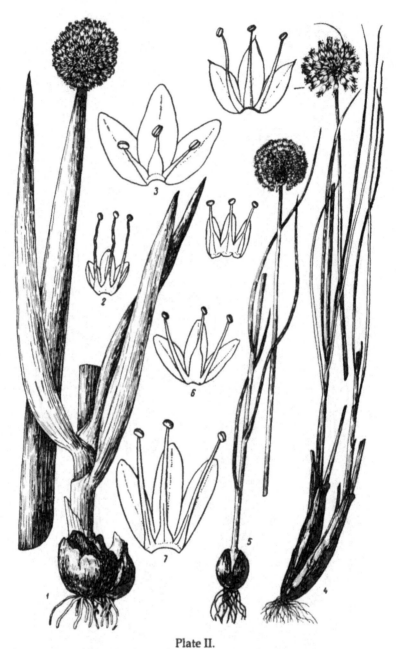

Plate II.

1—*Allium altaicum* Pall.; 2—*A. tianschanicum* Rupr. (flower parts); 3—*A. kansuense*
Rgl. (flower parts); 4—*A. petraeum* Kar. et Kir.; 5—*A. coeruleum* Pall.; 6—*A.
cyaneum* Rgl. (flower parts); 7—*A. platystylum* Rgl. (flower parts).

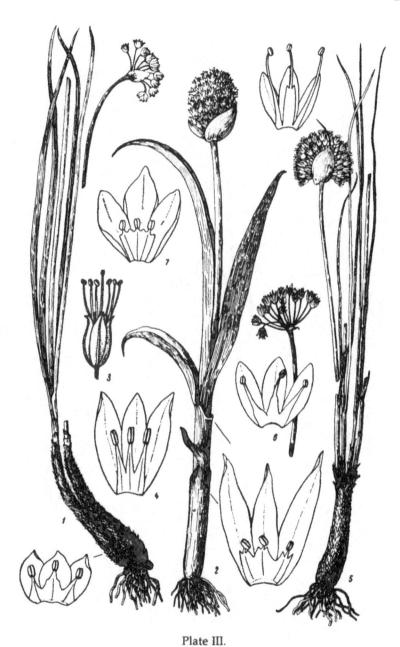

Plate III.
1—*Allium oreoprasum* Schrenk; 2—*A. semenovii* Rgl.; 3—*A. obliquum* L. (flower);
4—*A. chalcophengos* Airy-Shaw (flower parts); 5—*A. przewalskianum* Rgl.;
6—*A. odorum* L. (inflorescence and flower parts); 7—*A. fedtschenkoanum* Rgl.
(flower parts).

Plate IV.

1—Allium pallasii Murr.; *2—A. teretifolium* Rgl.; *3—A. sairamense* Rgl.; *4—A. glomeratum* Prokh.; *5—A. weschnjakowii* Rgl.; *6—A. robustum* Kar. et Kir. (flower parts); *7—A. decipiens* Fisch. ex Schult. et Schult. f. (flower parts); *8—A. anisopodum* Ledeb. (inflorescence and flower parts).

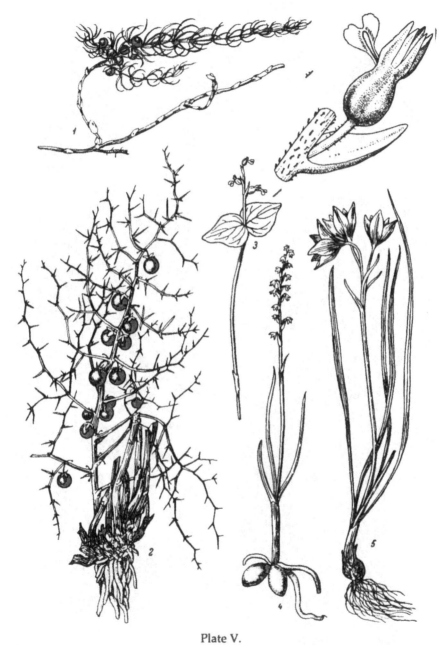

Plate V.
1—*Asparagus przewalskyi* Ivanova; 2—*A. gobicus* Ivanova; 3—*Listera tianschanica*
Grub.; 4—*Herminium alaschanicum* Maxim.; 5—*Lloydia tibetica* Baker.

154

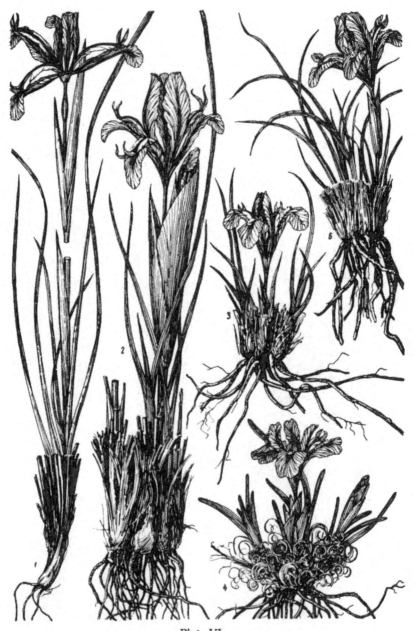

Plate VI.
1—*Iris maximowiczii* Grub.; 2—*I. bungei* Maxim.; 3—*I. potaninii* Maxim.;
4—*I. thoroldii* Baker; 5—*I. pandurata* Maxim.

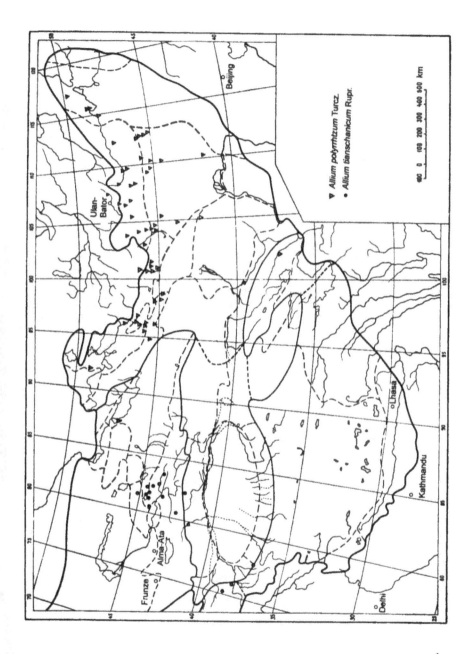

Map 1.

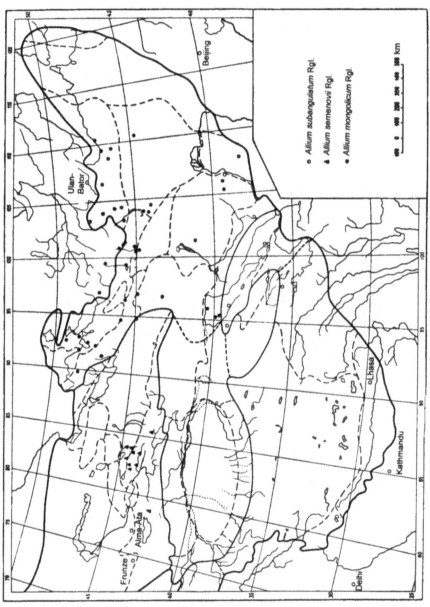

Map 2.

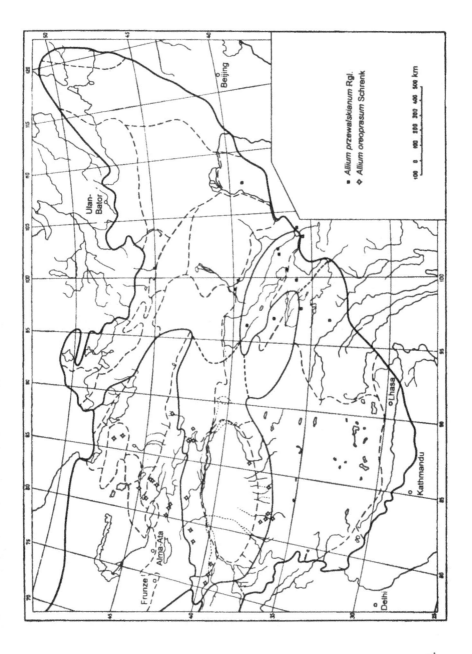

Map 3.

158

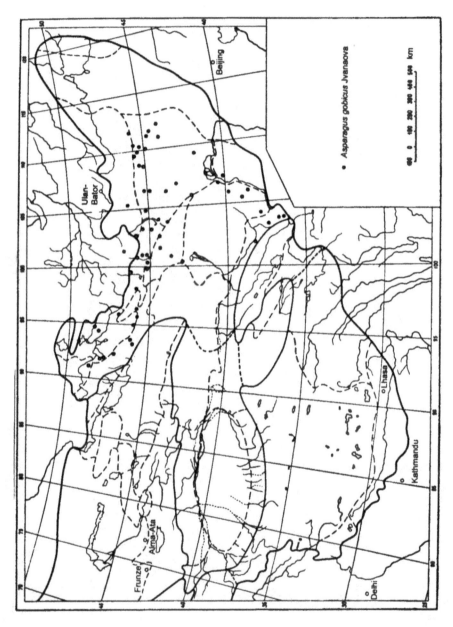

Map 4.

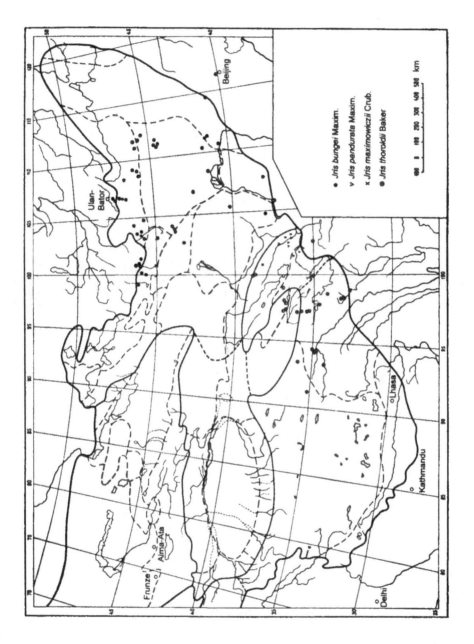

Map 5.

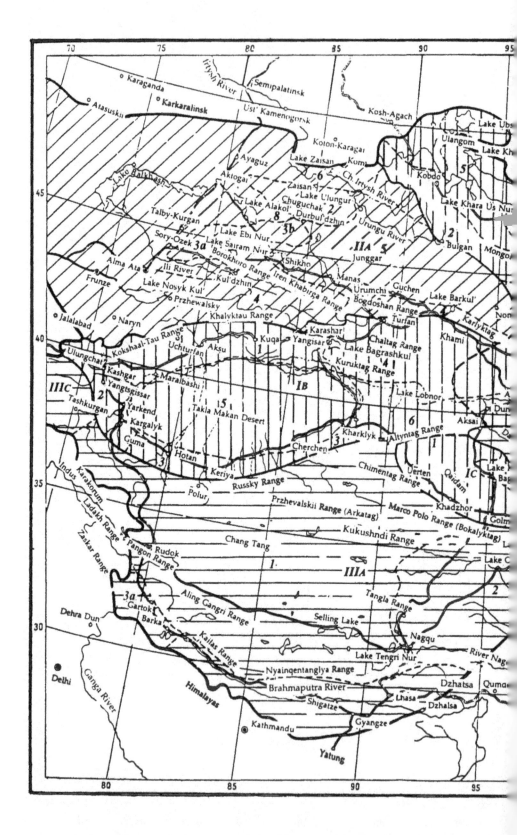

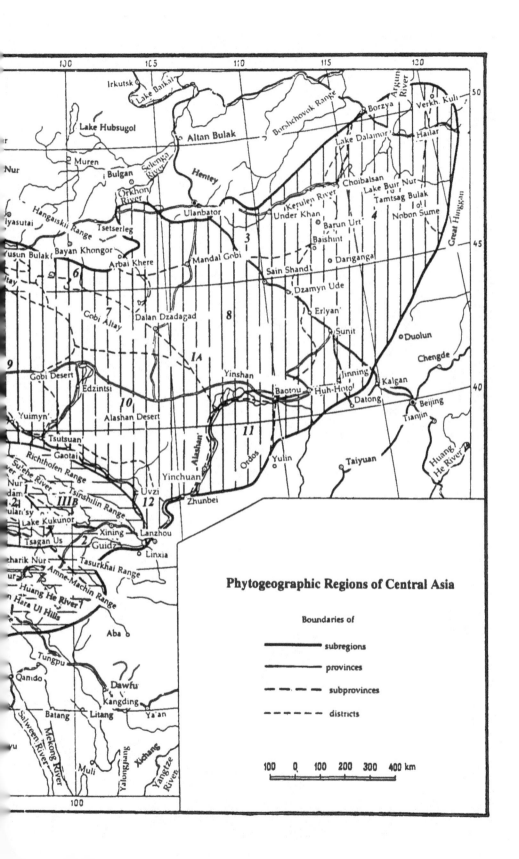

Phytogeographic Regions of Central Asia

Boundaries of

———————— subregions

——————— provinces

— · — · — subprovinces

— — — — districts

100   0   100   200   300   400 km

PHYTOGEOGRAPHIC DIVISION OF CENTRAL ASIA

(Central Asian Subregion of the Mediterranean)
(Revised)

In the geographic distribution of the species given in the chapter "Taxonomy", brackets [ ] indicate abbreviated names of regions; the absence of brackets signifies that the name of the region is shown in full.

I. MONGOLIAN PROVINCE (PROVINCIA MONGOLICA)

    IA. Subprovince Mongolia [IA. Mongolia] (Mongolia)

        Districts: 1. **Khobdo Basin [Khobd.]** (Chobdo)

                  2. **Mongolian Altay [Mong. Alt.]** (Altai mongolicus)

                  3. **Central Khalkha [Cent. Khalkha]** (Chalcha media)

                  4. **Eastern Mongolia [East. Mong]**
(Mongolia orientalis)
Region: a. Yinshan (Inschan)

                  5. **Great Lakes Depression [Depr. Lakes]** (Depressio lacerum majorum)

                  6. **Valley of Lakes [Val. Lakes]** (Vallis lacuum)

                  7. **Gobi Altay [Gob. Alt.]** (Altai gobicus)

                  8. **Eastern Gobi [East. Gobi]** (Gobi orientalis)

                  9. **Western Gobi [West. Gobi]** (Gobi occidentalis)

                10. **Alashan Gobi [Alash. Gobi]** (Gobi alaschanica)
Region: Alashan (Alaschan)

                11. **Ordos** (Ordos)

                12. **Khesi** (Chesi)

    IB. Subprovince Kashgar [IB. Kashgar] (Kaschgaria)

        Districts: 1. **Northern Kashgar [Nor.]** (Kaschgaria borealis)

                  2. **Western Kashgar [West.]** (Kaschgaria occidentalis)

                  3. **Southern Kashgar [South.]** (Kaschgaria australis)

                  4. **Eastern Kashgar [East.]** (Kaschgaria orientalis)
Regions: a. Turfan (Turfan)
b. Khami (Hami)

                5. **Takla-Makan** (Takla-Makan)

                6. **Lobnor Plain [Lobnor]** (Lob-nor)

    IC. Subprovince Qaidam [IC. Qaidam] (Tsaidam)

        Districts: 1. **Qaidam Plain [Plain]** (Tsaidam planitierus)

Regions: a. North-western [Nor.-west.]
(boreali-occidentalis)
              b. Southern [South.] (australis)
              c. Northern or Syrtyn Plain [Syrtyn]
(Sirtyn)
      2. **Montane Qaidam [Mount.]** (Tsaidam montanus)

II. JUNGGAR-TURANIAN PROVINCE (PROVINCIA SONGARICO-TURANICA)

IIA. Subprovince Junggar [IIA. Junggar] (Songaria)

Districts: 1. **Cis-Altay [Cis-Alt.]** (Paraaltai)
            2. **Tarbagatai [Tarb.]** (Tarbagatai)
            3. **Junggar Alatau [Jung. Alatau]** (Alatau cisiliensis)
            4. **Tien Shan** (Tianschan)
                Region: Kul'dzha (Kuldsha)
            5. **Junggar Gobi [Jung. Gobi]** (Gobi songarica)
                Regions: a. Central [Cent.] (centralis)
                          b. Northern [Nor.] (borealis)
                          c. North-western [Nor.-west.]
(boreali-occidentalis)
                          d. Western [West.] (occidentalis)
                          e. Southern [South.] (australis)
                          f. Eastern [East.] (orientalis)
             6. **Zaisan Depression [Zaisan]** (Depressio Zaissan)
            7. **Dzharkent Depression [Dzhark.]** (Depressio Dsharkent)
            8. **Balkhash-Alakul' Depression [Balkh.-Alak.]** (Depressio Balchasch-Alakul)

III. TIBETAN PROVINCE (PROVINCIA TIBETICA)

IIIA. Subprovince Qinghai [IIIA. Qinghai] (Tsinghai)

Districts: 1. **Nanshan** (Nanschan)
            2. **Amdo** (Amdo)

IIIB. Subprovince Tibet [IIIB. Tibet] (Tibet)

Districts: 1. **Chang Tang** (Czantan)
            2. **Weitzan** (Weitzan)
             3. **Southern Tibet [South.]** (Tibet australis)

IIIC. Subprovince Pamir [IIIC. Pamir] (Pamir)

The following regions (abbreviated) have been adopted for indicating the general distribution of the species: Aralo-Casp., Fore Balkh., Jung. Tarb., Nor. Tien Shan, Cent. Tien Shan, East. Pam.; Arct. (Europ., Asian), Europe, Mediterr., Balk.-Asia Minor, Fore Asia, Caucasus, Mid. Asia, West. Sib. (Altay), East. Sib. (Sayans), Far East, Nor. Mong. (Fore Hubs., Hent., Hang., Mong.-Daur., Fore Hing.), China (Altay, Dunbei, North, North-west, Cent., East, South-west, South, Hainan, Taiwan), Himalayas (west, east., Kashmir), Korean Peninsula, Japan, Indo-Mal., Nor. America, South America, Africa, Australia, New Zealand, Circumbor., Cosmopol.

INDEX OF LATIN NAMES OF PLANTS

INDEX OF PLANT DISTRIBUTION RANGES

INDEX OF PLANT DRAWINGS

Milton Keynes UK
Ingram Content Group UK Ltd.
UKHW020031071024
449327UK00032B/3012